13 FOODS THAT SHAPE OUR WORLD

13 FOODS THAT SHAPE OUR WORLD

How Our Hunger Has Changed
the Past, Present and Future

Alex Renton

BOOKS

BBC Books, an imprint of Ebury Publishing
20 Vauxhall Bridge Road,
London SW1V 2SA

BBC Books is part of the Penguin Random House group of companies
whose addresses can be found at global.penguinrandomhouse.com

This book is published to accompany the radio series entitled
The Food Programme broadcast on BBC Radio 4.

First published by BBC Books in 2022.

www.penguin.co.uk

A CIP catalogue record for this book is available from the British Library

ISBN 9781785947384

Commissioning Editor: Nell Warner
Project Editor: Liz Marvin
BBC Editor: Dimitri Houtart
Design and typesetting: seagull.net
Illustrator: Edward Bettison
Production: Antony Heller

Printed and bound in Great Britain by Clays Ltd, Elcograf S.p.A.

Penguin Random House is committed to a sustainable future for our
business, our readers and our planet. This book is made from
Forest Stewardship Council® certified paper.

CONTENTS

FOREWORD

BY SHEILA DILLON

The Food Programme set sail in September 1979 as a one-off series of six programmes on Radio 4. Journalist Derek Cooper, its first presenter, had lobbied the BBC for years to commission a series that would take food seriously. Finally, Radio 4 agreed. Derek later told the story of how, after the broadcasts started, he went with his producer to see the network's controller. They asked her if they could continue after the six episodes she'd commissioned. She was startled: 'But won't you have said everything there is to say about food by then?' she said. Derek persuaded her that there might be more to say and she reluctantly agreed they could carry on. Forty-three years later, it's still carrying on.

As the thirteen chapters of this book show, there's always more to say, taste and investigate. The food world is ever and fast changing and as a lens through which to understand the bigger world, food is unmatched. The history of the programme is the history of extraordinary change.

I first heard *The Food Programme* sometime in the mid-1980s. I was back from six years in New York where I'd been working as a journalist, for the final two years at the radical magazine *Food Monitor*. I didn't know about *The Food Programme*, I was just listening to Radio 4 – glad to be back in a country with great speech radio – but my attention was caught by a man's voice, rich and deep, describing from somewhere in France a celebration of local wines and British cheese.

Well, that was what it was meant to be, but the promised British cheeses hadn't arrived, so what was actually on offer to the gathered French winemakers and locals were two big blocks of unnamed cheddars, one white, one yellow – 'dyed yellow', said the voice. What a pity it was, he went on, because compared to the wines, they tasted of almost nothing. No one knew where they came from; they were obviously industrially produced and in no way represented Britain's great regional and farmhouse cheeses. Her Majesty's government's consul, who was hosting the event, was outraged and declared in fruity tones that they were 'jolly nice' and even better than the ones he ate at home. Derek very politely skewered HM's representative and cast a new light on the difference between industrial food production and the quality kind, and between a nation that sees good food as a pillar of its identity and one that doesn't. But it also showed the strength of a programme that could ask difficult questions, be journalistic and yet be based on pleasure.

That was the day I decided *The Food Programme* was where I needed to work. I bought a *Radio Times*, looked up the name of the producer and wrote to him. It took a few years and persistence, but I was eventually hired as the programme's reporter.

• • •

Outside Broadcasting House in London now there's a statue of George Orwell, author of *Animal Farm* and *1984*, and one-time BBC producer. He's poised as if on a soapbox ready to put the world to rights – fag in one hand, the other on his hip. On the wall beside it is carved an Orwell sentence: 'If liberty means anything at all, it means the right to tell people what they do not want to hear.'

In 1979, there weren't a lot of people who wanted to hear that food mattered. Even now, in government, the civil service and the big policy making organisations there still aren't. I know at first hand what difficulties the BBC's own news and current affairs division has in grasping that the production and consumption of food is actually as important as daily reports from the political centres. This is representative of a society where, for a long time, food has been seen as domestic and female, therefore trivial and ignorable. An idea embedded here since at least the beginning of the Industrial Revolution. The highest value that food has had in Britain is cheapness at the till. The idea of it as a civilising pleasure, the key to health and crucial for the way it connects us to nature, has never taken hold.

In the post-Second World War years, you could make a strong argument that there were many good reasons for food businesses and government to pursue an even more energetic cheap food policy. After the stringencies of rationing, which lasted in some foodstuffs until 1954, it seemed to make sense to industrialise the food system and bring down the cost of food. Britain was rebuilding, people were tired of food restrictions and ready for change.

Nobody challenged this model for a long time until, oddly, the Thatcher/Reagan years. That philosophy of public=bad/private=good saw the dismantling of controls on corporate mergers and takeovers and the general monetisation of the public sphere. Corporations got bigger. In the food world, the Milk Marketing Board, which had protected prices paid to dairy farmers, was killed off. Supermarkets, with their built-in incentives to buy wherever was cheapest, gradually took over retailing. Our food supply globalised.

A consequence was that wholesale markets went into free-fall. If you were a large-scale grower who didn't want to supply supermarkets because of the costs they imported (having to buy shelf space, pay for 2-for-1 offers, etc) there were suddenly very few options. As we showed on *The Food Programme* many times, a grower could be brought to the edge of bankruptcy in a world where the supermarket buyer had all the power and the growers had no contracts. In Cornwall, we saw acres of ready-to-be-harvested cauliflowers that had been declined by the supermarket they'd been grown for because they were cream coloured not white. They were going to be ploughed under because even Covent Garden, the biggest wholesale market in the UK, was no longer big enough to sell on this quantity of vegetables.

Meanwhile, the rise of out-of-town shopping centres centred around giant supermarkets sucked business from high streets: it was a rare butcher, baker, greengrocer or fishmonger who could compete on price or convenience.

It was into this world that *The Food Programme* was launched, to set about the task of taking food seriously and to look at the food system beyond the never-much-thought-about industrial model. Who was benefiting? What was happening to the quality of our food? Did it taste as good from a mega-factory as it did from a dairy or high-street baker? Where were the ingredients from? What additives were being used to cut the costs of production? What effect was that quality having on our health?

Those were key questions, but *The Food Programme* isn't a documentary series – the questions arose from the character of the man who launched the programme. Derek Cooper took food seriously, but for him it was also one of life's greatest pleasures. It's that never-broken link that's made the programme's reputation. For all its seriousness, its forty-three years are packed with joy, bringing alive, through food, people and cultures all over the world. We have reported from Sweden during the Santa Lucia feast and from the Rift Valley in Ethiopia, where after a devastating famine farmers stopped growing the hybrid maize as advised by development experts, returned to crop rotations and were now eating well and exporting their surplus beans, coffee and vegetables. In Tanzania, Dan Saladino joined a group of Hadza hunting and feasting. In Russia and Ukraine after the collapse of the Soviet Union, Derek Cooper and I talked and ate with bakers and market traders, and pensioners whose pensions were suddenly worth nothing in the gangsterland of that strange time. We've eaten the best tacos in Mexico City while researching the history of maize and Reestit mutton dried in open sheds in Shetland. Dan Saladino was in Venezuela reporting on a food system in crisis as the government edged near to collapse. In Chicago, producer Margaret Collins and I ate in the city's most starry restaurant with a group of exquisitely mannered and confident teenagers from the city's toughest schools as part of the chef's weekly mission to expand tastes beyond Coke and burgers. We've eaten with young refugees in Greece and enjoyed the best pizza with Dan's dad. In 2019, Dan and I were joined by two new presenters, brewer Jaega Wise and world-traveller and gardener, Leyla Kazim. They're already bringing fresh views and stories told through the lens of food and drink.

Our occasional *A Life through Food* series has brought us together with Paul McCartney, Madhur Jaffrey, Pret a Manger founder Julian Metcalfe, Prue Leith, Jamie Oliver, Nadiya Hussain, Nigella Lawson, Rick Stein and many more. Joy and journalism.

The programme's journalism got its biggest boost in early 1988 when we helped break the BSE story with Dylan Winter, reporter and producer on Radio 4's *Farming Today*. It soon became clear, though it was not admitted by the government for a long time, that mad cow disease was a catastrophe caused by the pursuit of ever-cheaper food. A catastrophe that was to cost the taxpayer multiple billions of pounds – the final total is still being calculated. The circumstances that led to BSE in our farm animals and the wide-reaching consequences are discussed in chapter twelve.

The BSE crisis was a tipping point in public trust of the government and experts. Importantly, it also showed that cheap food had a cost – in this case an astronomically high one. Up to then, only a few policy wonks and fewer journalists had questioned our cheap food supply. Now it is part of a bigger conversation and a spur to making food in a different way. Up rose the cheese makers, bakers, charcuterie makers, young butchers, growers, organic box schemes, farmers' markets, restaurants that name their suppliers ... all the elements of our food system we now take for granted.

Britain hasn't become a food paradise but change towards better quality has happened and continues because the questioning that was so rare in 1979 has become part of the food landscape. You don't have to be a *Food Programme* reporter to do it. If you're lucky and have an adequate income, you're now able to look at and use the food system in a very different way. Though nowadays, that is a bigger 'if'. Society is more unequal than it was in 1979 and with a less robust social safety net.

About fifty-five per cent of the food we eat in the UK in 2022 is defined as ultra-processed: made mostly from industrial derivatives of whole fats, starches and sugars, usually with the addition of artificial colours, flavours and stabilisers. We can measure the cost of this type of diet in the human misery of type 2 diabetes, heart disease, obesity, strokes and many types of cancer. The costs to the environment can be measured in rain forests chopped down to grow soya for cattle feed, in soils degraded by years of unbroken intense production of single crops, in the misery of animals reared intensively in factories, in methane and carbon emissions and the resulting rising global temperatures. As the cost of living continues to rise, food cheap at the till is still the national goal and the consequences of that are now beginning to overwhelm us.

Newspapers, radio and TV still talk of the 'need' for cheap food. Recently, I heard a news reporter interviewing James Rebanks, farmer and author of the bestselling *English Pastoral*. The reporter was critical because, he said, Rebanks appeared to be arguing that food should be more expensive when, in fact, we all know that 'we need cheap food to feed the poor'. Rebanks said, 'You don't fix poverty by making food cheaper, you fix poverty by redistributing income. What you've said is just propaganda.' The reporter quickly changed the subject.

But as Alex Renton shows in his exploration of thirteen foods, you don't get energy to keep questioning from fixating on the flaws. You get that by looking at the change makers. Which is why in 2000 *The Food Programme* and Radio 4 lobbied Prince Charles and Stephen Fry to help us launch the BBC Food and Farming Awards. The awards have just had their twentieth outing, rewarding people all over Britain – cooks working in institutions such as hospitals and schools, food and drink producers, street food vendors, retailers, entrepreneurs, markets, farmers – who are changing lives and economies around them through good food.

The awards are the BBC at its Reithian best: educating, informing and entertaining. The recordings we make on judging visits around the country and at the event itself produce great radio – and podcasts. My upright Yankee husband always has tears running down his cheeks as he listens. Even though I've heard it all on the road and on tape as we produce the programmes, I am moved all over again as I hear those voices go out into the world. As the BBC's director general Tim Davie said to the production team backstage at the 2021 awards, THIS is the beating heart of the BBC.

The Food Programme has been accused of bias over the years in favour of quality food and the people who produce it. The programme has always had to tread a fine line – giving industry its say because that is fair and because there is nothing at all wrong with making a profit. But we now live in a world where four corporations control ninety per cent of the global grain trade; four control just a little less of the meat industry, with just three companies dominating the production and sale of seeds and pesticides. In Britain, four supermarkets dominate food retailing. What this all means is that the global corporations that totally dominate our supermarket shelves have massive PR and advertising budgets to tell their version of the food story and protect their bottom lines. In 2017, the Obesity Health Alliance showed that junk food companies in the UK spent on advertising more than twenty-seven times the government's budget to promote healthy eating. We can't ignore that any more than we could if we were reporting on the oil and gas industries, which we now know have, for a long time, twisted the truth about climate change and so have shaped our response to the climate crisis.

The BBC's guidelines direct us to report stories of significance and to hold power to account. Like every other factual programme produced by the BBC, at *The Food Programme* we work hard to get at the truth with journalism that is fair and accurate. And sometimes that means telling people what they don't want to hear, just as Orwell said.

This book is an extension of that kind of journalism. I hope it brings you joy and new insights.

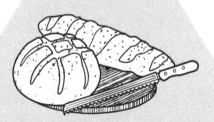

CHAPTER 1

FEED ME NOW AND EVERMORE
BREAD

'By reimagining our daily bread, Andrew Whitley is convinced it's possible to build a food system fit for the planet and fit for us.'

Dan Saladino on the veteran real bread campaigner Andrew Whitley, *The Food Programme*, 2021

Bread is civilisation: an artefact whose making demands science, agriculture, art and, most of all, cooperation. Its simple inputs – flour, yeast, salt, water and heat – are each staples of human life, and, except water, products of acquired knowledge. You might add to the mix one more crucial thing: time, carefully portioned. Yeast needs time to grow, dough needs time to prove, bread needs time to bake and cool.

Bread may be one of the oldest things we make. Like other animals, we have long eaten the seeds of grasses; archaeological finds show that we began grinding these between stones before we appear to have begun taming animals for food. This rudimentary flour was used for porridges. In the mix, bacteria thrived, giving off carbon dioxide, which add taste and volume to the seed sludge.

The first meal of bread we know of, currently, was eaten in a remote area of what is now Jordan's eastern desert, towards its border with Iraq, about 14,400 years ago. In 2017, charred food remains were found in a fireplace at a site called Shubayqa 1. Under an electron microscope, they proved to be the remains of grains that had, as the writer Rob Penn puts it in his book *Slow Rise*, 'been threshed, winnowed, milled and possibly sieved, before being mixed with water into a form of dough and cooked ... this is the beginning of bread'.

From there, a human history with bread as symbol and staple begins. Bread is the 'staff of life' in the Bible and 'bread of heaven' – in the form of *manna* – is sent from the skies to feed the wandering Israelites. Bread and power go hand-in-hand: as with other foods in this book, finding a way to process food energy so it can be stored, transported and traded has been a foundation of great civilisations.

Bread has kept the poor from rising against the rich. In ancient Rome, free grain was distributed by the ruling class to the *plebs*: 'bread and circuses', as the satirist Juvenal put it, was the recipe for peace in the city. Subsidising bread and its key ingredient, flour, has been necessary in many societies ever since, with dire consequences for the ruling classes should they fail. 'Let them eat cake,' the eighteenth-century French queen Marie Antoinette is supposed to have answered when told the poor could not afford bread. The ordinary people of France removed her head. (The word she used in French was *brioche*, which – with its eggs and sugar – is more expensive than plain baguette: a cruel joke, if she ever said it.)

Londoners rioted over bread prices several times in the eighteenth and nineteenth centuries. Political reforms followed. Bread remains as symbolic and political as ever: when world wheat prices soared after 2008, the cost of bread in some countries rose by thirty per cent or more. This became an issue in the popular protests around North Africa and the Middle East known as the Arab Spring. When demonstrators took to the streets of Cairo in

2011, one of their chants was a demand for '*Aish, Horreya, Adala Egtema'eya*' – bread, freedom, social justice: the protests ended in the fall of the government.

The quest for grain for bread has removed forests and reshaped landscapes, from America's prairies to the great grasslands of Russia. But for all the damage done, we have some of the greatest – and most joyous – products of human ingenuity. Focaccia, challah, baguette, pita, pretzel, sourdough, arepa, lavash, bagel, ciabatta, njera, simit, pumpernickel, brioche, the crumpet. (Chapati, tortillas, scones and other breads that are not leavened don't make it on – the usual definition of bread is that it has to be made with yeast.)

4

'The twentieth century was the first since civilization began when the majority of us were not steeped in growing and harvesting cereals, then threshing, milling, baking them into bread.'

Robert Penn, *Slow Rise* (2021)

Bread is togetherness. Company, a companion – from the Latin and old French – is someone with whom we eat bread. The ancient Greeks called the Egyptians 'bread eaters' – the modern word for it in Egypt today, *aish*, also means 'life'. It seems fair to assume that ever since humans first baked bread it has been a metaphor for and a token of comfort, money, partnership and community.

But in the twentieth century bread changed radically as time – that crucial sixth ingredient – became more expensive. We started looking at the hours spent bread-making as a waste, a chore. We invented more and more processes to bypass the natural process and make bread more profitable. This meant that bread could be made quicker, at a large scale and lasted longer. But the result was that we lost our connection with the ingredients, the slow rhythms and rituals of crafting a good loaf.

Bread is now getting better – for those who can afford it. For example, in Edinburgh in 2021, there are a dozen or more small-scale artisanal bakers around the city; my family gets an organic sourdough loaf delivered to our doorstep once a week by the city's Company Bakery. But that costs more than three times the price of supermarket sliced wholemeal, making it beyond the means of many.

BAD BREAD

'Why do the rich in so many countries eat such bad bread?' asks food historian Bee Wilson in her polemical history, *The Way We Eat Now* (2019). The poor, she goes on, have less choice in the matter. It is clear that in Britain the ordinary loaf started to deteriorate nearly two centuries ago. As Britain industrialised, busy urban workers looked to others to make their basic foods.

Some workers' housing was built without a kitchen. But this shift away from home-cooking enabled fraud. In the cities in the 1850s, bakers were habitually adding alum, which is toxic, along with ground bones and chalk – these bulked out the flour and made loaves whiter. Already brown bread was looked down upon as the food of peasants. There was surprisingly little protest. Bee Wilson writes in her chronicle of food fraud, *Swindled,* that 'a significant strand of opinion reckoned that a bit of harmless adulteration was defensible as good for trade': 'buyer beware' was a fair rule and politicians resisted calls for regulation.

In France, by contrast, strict rules about the content and making of bread were enforced from the time of the French revolution. In 1793, the new Republic's government stated: 'It will no longer make a bread of wheat for the rich and a bread of bran for the poor. All bakers will be held, under the penalty of imprisonment, to make only one type of bread: the Bread of Equality.' This, and subsequent strict standards brought in during the nineteenth century, ensured a national reputation for great bread that, though factory-made loaves are found in France too, lasts until today.

In Britain, an Adulteration Act was eventually passed in 1860, marking one of the first times that the government intervened to regulate food. But by then, bread, a key food for much of society, was doing active harm, chiefly because it was all made from bleached white flour and so lacked almost all the nutrients and vitamins that the grain provides. In 1910, one appalled observer of working-class south London, socialist and suffragette Maud Pember-Reeves, reported that white bread was most, if not all of what the children she inspected ate.

WHITE AND SPEEDY

Quite when the British started choosing white, bleached flour above all else is not certain. The reason probably lies with the same prejudice that keeps Asia eating white rice: brown, less milled grain is a stigma of poverty. The 1890 edition of Isabella Beeton's immensely successful *Every Day Cookery* has a recipe for 'Good Homemade Bread', which does not specify a type or colour of flour – she must have assumed her audience used white. Neither does *A Plain Cookery Book for the Working Classes*, a bestseller by Queen Victoria's personal chef, Charles Elmé Francatelli, published in 1852. But his recipe 'Bread, how to bake your own' seems written for people who did not normally do such a thing.

That's understandable: heating an oven to over 200°C to bake bread is expensive in fuel. Buying your bread from the bakery – even if it was spiked with alum, ground bones and bleach – was a better option for urban people.

Toast sandwich – the cheapest meal ever

This recipe is from Isabella Beeton's other bestseller, *Mrs Beeton's Book of Household Management*, first published in 1861. 'Place a very thin piece of cold toast between two slices of thin bread-and-butter in the form of a sandwich, adding a seasoning of pepper and salt,' she wrote. The Royal Society of Chemistry surveyed parsimonious British recipes and declared this the cheapest lunch ever published, a 330-calorie snack costing just 7.5p a helping today.

For many women, being relieved of the ancient chore of making bread was a huge boon. Mrs Beeton praises a Dr Dauglish and his new technique for 'aerated bread', machine-made in factories by pumping carbon dioxide into the dough. These and more industrial changes will, she hoped, 'emancipate the housewife and the professional baker from a large amount of labour'.

Fuel is cheaper today but home baking remains expensive in another way. Sarah Bridle calculates in her *Food and Climate Change* (2021) that making a loaf of your own bread is ten times as costly in terms of emissions of climate-affecting gases as buying one from a shop. The cost is all in the heating of the oven.

BREAD FOR A BETTER BRITON

8

Maud Pember-Reeves's reports on bread-fuelled life in working-class Lambeth got an audience. The following year, the *Daily Mail* – then as now one of Britain's most powerful newspapers – launched a 'Standard Bread' campaign. It was backed by the King George V's doctor and a celebrity MP Sir Oswald Mosley (the father of the future British Fascist leader).

The newspaper reported that since bread, almost always white, made up forty per cent of the diet of the poor, it might well be to blame for both their 'degeneracy' and 'the decline in the national physique'. The last was a much-discussed problem. Poor diet appeared to be shrinking the nation. The British Army had had to reduce its minimum height requirement for recruits, down six inches to five foot in 1902. Nearly half of all those who qualified still had to be rejected because of poor health and bad teeth. The *Mail* made much of this.

Something had to be done to restore 'the cream-coloured loaf', said the paper and 'oust the pasty-white usurper'. 'The delicious old farmhouse and cottage bread made the bone and sinew of Englishmen and the beauty of English women famous before the craze for white starch bread began,' it went on. A prize of £600 (£55,000 in today's terms) was offered for a better recipe.

It would be more than a decade until an understanding of vitamins led to the first steps in modern nutrition science, but the *Mail* had put its finger on the problem. The white bread that was then ubiquitous would certainly have contributed to a nationwide vitamin B deficiency, since the germ of the wheat was discarded during the milling process – and rickets and other related diseases were rife in poorer Edwardian Britain. Unregulated chlorine bleaching of flour would also have added toxins to the daily bread.

By late 1911, the newspaper was able to proclaim a triumph: 'A Change in the Food of the Nation' had been achieved. Countless homes were eating the *Mail*'s Standard Bread, the paper claimed, made from unadulterated flour and at least eighty per cent of the germ of the wheat. A loaf was seen to be delivered daily to Buckingham Palace. But Lord Northcliffe, the owner of the *Mail*, failed to persuade the government to legislate for the bread. Some of his detractors said he'd only started the campaign to prove how powerful was his newspaper.

In the Second World War, white bread was banned in Britain partly for the same health-related reasons (though with no mention of degeneracy, this time). A nation under rationing and on the brink of starvation could not afford to waste so much goodness from the flour that it largely imported across the Atlantic. In America, the same rule was brought in in 1943, including a ban on sliced bread because it went off more quickly. But the public rebelled and pre-sliced bread, which eighty per cent of the nation bought after it was introduced in the 1930s, reappeared on the shelves.

When white bread was permitted again in Britain, in 1947, government nutritionists regretted it. At this point, the country, especially its children, was healthier in almost every respect than they had been in 1939 – and better bread was part of the reason why. White bread returned to rule again – although from the 1940s, home baking recipes did occasionally suggest that 'wheaten' (wholemeal) flour could be used in place of white. Though who actually baked their own bread in post-war Britain? People considered oddballs – vegetarians, anti-modernists and those with too much time on their hands.

10

'It's a nice loaf to look at … it suits the housewife who hasn't the energy and in many cases, hasn't even a knife to cut bread.'

A 1950s baker contemplating the mass-produced, pre-sliced loaf, BBC Archive

Bread in the 1970s

'My brother and I ate loaves of Mother's Pride or Sunblest.
Occasionally we ate an entire loaf at a sitting ... I was briefly
captivated by Ridley Scott's famous 1973 TV commercial for Hovis
– a bread delivery boy pushing a bicycle up a cobbled lane to
Dvořák's New World Symphony – but I never ate the brown bread
it advertised. It was nutrient-stripped white – white enough to blind
you – or nothing at all: white toast and honey for breakfast; white
bread to mop up liver and bacon stew at lunchtime; white toast
with Marmite and baked beans in front of *Grange Hill* for tea.'

Robert Penn, *Slow Rise* (2021)

There was a rebel hold-out, though. In 1954, Doris Grant published *Dear Housewives,* and for all its modest (and old-fashioned) title, it was a powerful tract for better food for ordinary British families. She was already famous for her espousal of the first balanced diet, the Hay System, her campaigning against rapacious food companies, and a delicious, wholemeal bread to make at home – the Grant Loaf.

'I remember the sensation this caused when it came out just after the war,' wrote one of the next generation's campaigners for real food, Jane Grigson, in 1974. 'Here was the real thing, and it took only a few minutes to make because no kneading was required.' Grant's 1954 book weighed in against 'Baker's Poison', with its 'witches brew' of additives, its lack of vitamins, its short shelf life and its tastelessness. She noted that if she put crumbs from her home-baked bread and some from a bought white loaf on the window sill, the brown crumbs were taken by the birds first. And her Labrador wouldn't touch the white bread.

Nonetheless, as Doris Grant was writing, the bread industry was coming up with its most efficient idea yet. Put together by scientists at a baking industry research institute outside London, the Chorleywood Baking Process (CBP) would soon be producing most of the bread in Britain and in many other countries.

In a form, the CBP is still responsible for ninety per cent of the bread we eat today. High-speed mixing and baking and low protein wheat flour is bolstered by a huge range of processing aids, 'yeast improvers', preservatives, emulsifiers and other chemicals, with high amounts of salt and sugar often added, too. Launched in 1961, it produced a softer, lighter bread than ever seen before – quickly, consistently and very cheaply. Not only did it bring down the price of a loaf, the bread also kept longer. This helped the increasing numbers of women entering the workforce, who no longer had time to shop every day.

Bread additives and improvers found in a standard supermarket loaf

Hard fats, soya flour, acetic acid, powdered gluten, reducing agent (l-cysteine hydrochloride), flour treatment agent (ascorbic acid), emulsifiers (mono- and diglycerides of fatty acids, sodium stearoyl-2-lactylate, glycerol monostearate lecithins and others), bleaching agents, preservatives (calcium propionate) and enzymes.

To some, it was much more attractive – particularly for its lack of any crust worth mentioning. But to others, it was a soft, moist, tasteless cotton wool. 'When a mollusc of gummy paste clamps to the roof of your mouth like a limpet to a rock … you know you are eating CBP bread,' writes Robert Penn in *Slow Rise*. Elizabeth David, a warrior for real food in the mid-twentieth century, said, 'A technological triumph factory bread may be. Taste it has none. Should it be called bread?' America's equivalent of Elizabeth David, the TV chef and writer Julia Child, famously asked: 'How can a nation be called great when its bread tastes like Kleenex?'

The baking industry has lobbied hard and successfully to ensure that many of the additives it uses don't have to be included in the ingredients list on the packet. Its argument is that since these 'processing aids' and 'yeast improvers' help with the rise of the dough and disappear during the baking, they are not actually present in the finished bread. But they do affect the bread in many different ways and, as has become clear, they affect our bodies too.

Doris Grant's concerns over the effects on our health of factory-made bread in 1954 – drawn from her observations at the dawn of nutrition science – are pretty much the same as those that worry many of us now. Deficiencies in vitamin B complex and vitamin E are still an issue; rickets has not gone from Britain, even among wealthier households, and lack of calcium may play a part in poorer Britain's world-beating bad teeth problem. Foretelling today's concerns, Grant called sugar 'public enemy no. 2': these days, supermarket bread may contain as much as three grams of sugar per slice.

The residues left by bleaching to make flour white are less of an issue, though this process does still happen. If white flour is not labelled 'unbleached', it will have been chemically aged to whiten it – a process that can happen naturally, but it is slow.

Perhaps most significant of all, Grant put her finger on an issue that affects almost all of us who think about and eat bread: wholemeal, with its fibre, makes our gut work better. 'Vitamin B strengthens the colon,' she opined – as a former sufferer of constipation herself. She was right, and more: it plays a role in reducing the risk of bowel and colon cancer, another illness whose rise in recent years has not yet been explained, though it has been suggested that the additives and lack of fibre in highly processed foods such as CBP bread are among the causes. To be fair, the bread industry has reacted to criticism over excessive additives. 'British bread now has the lowest salt content of any in Europe,' says Gordon Polson, head of the Federation of Bakers. The added sugar content (some sugar is naturally present) of ordinary bread also appears to be declining, though research across all brands is not available.

All the same, a century after Lord Northcliffe demonstrated his *Daily Mail*'s muscle with Standard Bread, Real Bread Campaign co-founder Andrew Whitley said that bread may well be worse than it was then. 'I think Lord Northcliffe would turn in his grave if he could see the state of factory loaves today … We want the government to give bread the same sort of protection as butter, so anyone wanting to throw in additives would have to come up with another name for it.'

RAISING THE DOUGH

The Grant Loaf was far healthier than standard white, and tastier. It was easy to make because it did still use one key modern ingredient:

a teaspoonful of instant 'live' yeast (7g for 450g of flour). Doris Grant puts its leavening and rising process at a mere twenty to thirty minutes, rather than one to three hours for a kneaded loaf, or twelve hours or more for a traditional sourdough. Although 'real bread' makers might turn up their noses at using a yeast starter they had not nurtured and grown themselves.

Yeast works by reacting with the flour and water in the dough, producing lactic bacilli that help make the bread tasty and digestible and, crucially, the gas carbon dioxide. These bubbles help raise the dough and stretch the gluten.

Done traditionally, it's not a fast process – from the first feeding of the starter, it may be twenty-four hours before you can take your loaf out of the oven. Sourdough bread – probably the first our Neolithic ancestors made – may be left to 'prove' for four hours or more, and then again overnight in the fridge. It isn't surprising that since the nineteenth century, commercial bakers have looked for shortcuts.

Ready yeast had long been available from breweries. In its natural state, it is a thick soup, bubbling and clearly alive. But it is not stable. By the late nineteenth century, manufacturers were extracting the moisture from the yeast culture, producing cakes that could be kept and sold. From that came instant yeast and then 'rapid-rise', which does not need to be rehydrated, and produces much more carbon dioxide. Standard bread using the Chorleywood process and enzyme-laced yeast can be made in an hour flat. Nowadays, CBP loaves are often frozen after they are proved and shipped to supermarkets for baking in store, enticing customers with that fresh bread smell.

But Andrew Whitley argues in his ground-breaking 2006 book *Bread Matters* – both a campaigner's *cri de coeur* and a great bread-maker's manual – that bread made slowly has a much better flavour and texture, as well as containing a host of nutrients produced by natural fermentation. Some quick recipes for home breadmaking machines call for 15g of instant yeast to 400g of flour (more than twice as much as in the Grant Loaf recipe). However, undigested yeast causes a range of problems in the human gut.

We eat far less bread that we used to – our consumption is down about fifty per cent since the 1950s and, according to the Federation of Bakers, it is still decreasing. Today in Britain, we eat about forty-three loaves per person, or 30kg each, a year: a century ago it was 150kg. Seventy-five per cent of bread sold is white. There are many reasons for this decline, including the rise of other carbohydrates. British families used to have a loaf of bread on the table at most meals: now we use rice and pasta. But this has not been matched by any radical changes in our diet to replace the beneficial elements present in real bread. Fruit and vegetable consumption has risen, but not in any way that makes up for those missing vitamins and fibre. The Federation of Bakers' Gordon Polson makes the point that, even though the British eat mainly white bread, it is still responsible for seventeen per cent of the fibre in the average person's diet.

Growing awareness of the importance of fibre in our diet has prompted the development of products such as 50/50 loaves made with part wholemeal, browner breads decorated with whole grains and even loaves posing as sourdough but in fact made with added yeast. Speaking to a bread industry representative in 2007, Sheila Dillon suggested, 'A Martian coming down here would say, "Why not just make variations of wholewheat?"' His response was that he – and the nation – liked white.

While many Britons stuck at home under lockdown in 2020 and 2021 took to making their own bread, Gordon Polson says that, looking at the whole market, white bread is as popular as ever today. Many of the artisanal breads that families may buy as a treat are still made with white flour. Many prefer it toasted or as a bacon sandwich. The modern white bread that most of us buy is, it could be argued, a vehicle for other flavours, rather than a food in its own right.

GLUTEN PHOBIA

By the early 2000s, the protests of the real bread movement were getting traction among the British public, and bakers, traditional and big-scale, were taking notice. But then a new worry emerged, with suspicions falling on the key protein in bread flour, gluten. As with other fashionable approaches to worries about health and lifestyle, the scientific basis for drastic alterations to diet on grounds of 'gluten intolerance' was shaky. In fact, some researchers have concluded that people should not be encouraged to avoid gluten unless they have coeliac disease because their overall diets would deteriorate.*

Nevertheless, fuelled by some health writers and then by social media influencers, 'gluten-free' became the biggest health fad since 'low carb' a few years earlier. In 2013, a poll found that twenty-nine per cent of Americans were trying to cut down on consuming gluten, or avoid it completely).† The story was similar in Europe – according to market research company Mintel, in 2018, fifteen per

---

* Kmietowicz, Zosia, 'Gluten-free diet is not recommended for people without celiac disease' *British Medical Journal* 2017; 357
† 'Gluten goodbye', npr.org, 9 March 2013

cent of British households were avoiding gluten and wheat. The anti-gluten craze was widely mocked: in the cartoon show *South Park*, the entire town went gluten-free, believing that gluten made penises fall off.

Nonetheless, by 2018, 'gluten-free' was available in most supermarkets. It had become a major line in what food marketing calls the 'free from' and 'better for you' sector, one of the most lucrative areas for new food products. According to a 2020 report by the charity Coeliac UK, the gluten-free market in the UK was up from £93 million in 2009 to an estimated £394 million in 2018. The label sprouted on dozens of products from beer – which does contain gluten – to chocolate and potato crisps, which are naturally gluten-free. This new obsession was not good news for artisanal bakers, who had begun to hope they were winning the long battle to restore real bread to its rightful place.

The demonising of a substance key to one of humanity's most ancient staple foods never made much sense. What is true is that the demands of modern fast-process bread baking has led to wheat strains being developed to contain more gluten. The industry has also been extracting gluten from wheat and adding it to fast-processed bread in order to give it texture. But bread made in the time-honoured way, from traditional flour, has the same gluten content it always had.

As a result, interest has grown in rediscovering some of the ancient varieties of seeds and plants used for flour: *The Food Programme* has highlighted organisations like The Grain Lab, which holds annual get-togethers for farmers, millers and bakers exploring heritage varieties and techniques. Baker and author Victoria Kimbell encourages home sourdough makers to try replacing twenty per cent of their usual flour with traditional wheats like emmer, rivet, einkorn and khorasan.

Gluten-free bread, pasta and pizza are not easy to make. It is possible, by using a range of chemicals and fats, to reproduce the spring usually provided by gluten from grain: a complex and expensive process. But consumers were willing to pay much more – a loaf of gluten-free white bread at Tesco costs £1.80 at time of writing, as opposed to 59p for its standard own-brand white bread.

The bread is expensive in part because new technology had to be developed to make it, says Gordon Polson of the Federation of Bakers. 'It's not really bread in the strict sense. Actually producing something that the consumer would want was technically extremely challenging and required a lot of investment.' Besides, if you remove the key element of a loaf's engineering, extra salt, fat and sugar are likely to be used to make a gluten-free loaf work. Andrew Whitley, who does provide some gluten-free recipes for bread, cautions, 'Wheat gluten is unique. You can't mimic it except by using strange additives that aren't food.'

The gluten-free trend continues today. It has clearly benefited the one per cent or so of the population who do suffer coeliac disease – there is more choice for them and better understanding of their serious condition. But no one has yet been able to nail down what has caused the undoubted rise in reporting of coeliac disease – or in irritable bowel issues, asthma and other symptoms sufferers were told to blame on gluten sensitivity.

Science, as so often is the case in food and health when big money is involved, seems unable to find a consensus. But a significant number of studies have found the links between these issues and traditional gluten use in food is tenuous and suggest that many problems may have different dietary origins.

There is some evidence that people who believe they have a problem with gluten react well if they change to less processed, slow-proven bread, especially sourdoughs.* And it may be that the problems in the modern gut have been caused, as Doris Grant said all those years ago, by the deficiencies and the artificial additives of factory bread.

A GOLDEN AGE OF BREAD?

Nathan Myhrvold, the ex-Microsoft millionaire who has spent part of his fortune investigating cooking processes in a laboratory, has published a five-volume, full-colour, big-format account of his explorations of bread baking. (As part of their investigation, 36,000 loaves were baked by him and his sixty chef-researchers.) *Modernist Bread* is pitched at a select audience – the book sells for £400.

No one could have more respect for traditional approaches than Myhrvold, a Minnesota wheat farmer's grandson. But he, throws out some time-honoured lore that he sees as myth. Talking to Dan Saladino on a 2018 episode of *The Food Programme*, Myhrvold discusses how the real bread movement began in the 1970s in France and with 'a bunch of hippies in northern California'.

He went on to say: 'They looked at the history of bread. They said we're ruining bread with this supermarket stuff; let's reach to the past and we can make really good bread. Over time, people kept

---

* See citations in www.theguardian.com/lifeandstyle/2016/mar/23/sourdough-bread-gluten-intolerance-food-health-celiac-disease

reaching a little further back. The way to one-up your competitor was to say, "You use a gas-fired oven? I use wood." Then, that guy'd say, "You buy your flour? I make my flour!" There are some good things that can come of that, but what's next? Stone tools? At what point are you actually making the bread better?'

The golden age of bread is not in the past, it is now, insists Myhrvold. The return to basic principles comes about through consumers' and bakers' objections to industrialised bread, but that does not mean we have to reject every modern or technological discovery. As Myhrvold sees it, artisanal bread-makers are foolishly snooty about 'no-knead bread', which has been around as a labour-saving method since the 1940s – as we saw, Doris Grant boasts how little kneading her loaf requires.

'Kneading is a fraud,' says Myhrvold. 'It doesn't do what is promised. You can make great bread without kneading.' The point is that, with the right yeast properly prepared, with the water and flour allowed time to 'autolyse', the very action of fermenting over eight hours will do all the stretching the dough requires. Adding a drop or two of acidic fruit juice helps, too.

The Myhrvold lab made several other significant discoveries. One was a natural way to combat the problem of the dryness and heaviness of wholemeal loaves; they tackle this by simply starting the bread with whiter flour and adding bran and wheatgerm later when the gluten has started stretching. He also gives precise instructions for a problem all home bakers face: what to do when you've over-proofed a dough and it collapses. The problem is that the gluten has stretched too far because of too much carbon dioxide from the fermentation: the whole structure is impossibly weak.

'Conventional wisdom holds that overproofed doughs are irretrievably damaged and should be thrown away,' Myhrvold writes. 'Our experiments found just the opposite. In fact, we were able to resuscitate the same batch of dough up to ten times before it suffered any serious loss in quality.' The answer is simple: collapse the loaf completely, shape it and start the proving process again. Though this won't work for sourdough, where the gluten structure is tougher and acid levels high, he warns, unless you spot the fault early in the proving.

After all that eating and experimenting – 'I put on about 15lb myself,' giggles Myhrvold – what one bread would he choose? The man who has probably tried more breads than anyone else on the planet picks a bread invented by his collaborator, the pastry chef Francisco Migoya: a chocolate and cherry sourdough. 'It's not a sweet bread, but the dark chocolate, and then these cherries … Boy! Words fail me.' The recipe is on the Modernist Cuisine website.

TWENTY-FIRST CENTURY PRESSURES

In terms of climate-changing gas emissions, the biggest impact in most homes comes from using the oven. Used regularly, it may be responsible for thirty per cent of emissions from the average household, far more than heating water.* There are ways to mitigate some of the impact: using the fan to reduce cooking temperature and limiting the time preheating an oven, for a start.

---

* See S.J. Bridle, *Food and Climate Change without the Hot Air* (UIT Cambridge, 2020), and www.bbc.co.uk/food/articles/cooking_carbon_footprint

But it is an unavoidable truth that mass, commercially baked bread is far more efficient than each of us turning on the oven to 240°C to bake a couple of loaves at home. If we shared an oven with our neighbours, as communities once did, the figures would look much better. The other course is to buy from the new breed of genuine high-street bakers who, with the support of organisations like the Real Bread Campaign, have done much to restore proper bread to those who have the means to go and find it. Though this is of course the expensive option – a loaf of sourdough from a local baker regularly comes in at around £4 or more – and is not available to everyone.

During the twenty-first century, the cereals we use to make our bread and the people who farm it will be faced with greater and greater challenges because of shifting weather patterns. Wheat provides twenty per cent of all human nutrition, and ninety-five per cent of the three-quarters of a billion tonnes of it grown annually for food worldwide is made into bread. Britain imports up to twenty per cent of the wheat it uses annually. We are quite dependent on the Canadian crop for bread, since it is what's called 'stronger', meaning higher in protein – the gluten that gives bread its structure. Because it produces so much high quality wheat, Canada fills in when British harvests are disappointing. But with climate change affecting crop patterns and yields across North America's great plains, this century-old relationship is now less secure.

Wheat shortages often cause crises in poorer countries and we would be unwise to think ourselves immune: in the post-Second World War period, Britain and other north European nations came so close to famine because of crop failures that the US, which was shipping in wheat to help us, began a public campaign to ask

American consumers to cut their bread eating by fifteen per cent. Poor harvest predictions in the biggest wheat-growing countries do still affect global stocks and prices. But today wheat prices are also controlled by markets and dominant, profit-focused multinational corporations, rather than by governments. We may all be more at risk of shocks from the climate than ever before.

How will we fare in the coming decades? Growing the grain that we use for bread in the UK produces some two million tonnes of greenhouse gas emissions each year. That sounds truly frightening but drinking one glass of milk (which, if farmed in Europe, produces 250g of carbon and other gas emissions*) results in nearly ten times the emissions from eating two slices of bread. At present, we feed twenty per cent of all world grain crops to farm animals – something that will have to change as the growing world population puts more and more pressure on global food production.

If people who can afford to do so accepted that bread, like so many other staples, is worth paying more for, that would begin to address the environmental issues associated with the farming of wheat, like the mass use of fossil-fuel based fertiliser, for a start. And bread, like so many staples in this book, is cheaper in real terms than at most times in modern history. In 1931, a standard 4lb (1.8kg) white loaf cost 7p in Britain – the equivalent of £5.58 today, in terms of the change in average wages. Two 800g standard white loaves at Tesco cost £1.19 in 2021.

Speaking to *The Food Programme*, Nathan Myhrvold quoted a United States Department of Agriculture study into all the costs

---

* S.J. Bridle, *Food and Climate Change without the Hot Air* (UIT Cambridge, 2020). 1g milk produces 2g carbon. Figures based on the European dairy industry, which is more efficient in climate change terms than others.

and revenues in a loaf of bread: 'The farmer gets five cents, the
smallest wedge in the pie chart. The plastic bag costs as much as the
grain did. Advertising costs more than the grain. Transportation,
more … We the consumers bear some responsibility. We're not
willing to pay more for it, which we are with coffee, with wine, with
chocolate …' He's echoing many other bakers when he says that
societies get the bread they deserve.

CHAPTER 2

ROCK OF AGES

SALT

'The book's title is to counter the demonisation of salt at the moment. Everybody who cooks and eats knows that salt is essential. You don't have to have too much but if you don't have enough salt you start to teeter towards blandness. And blandness is bad.'

Chef Shaun Hill, author of *Salt Is Essential, The Food Programme*, 2018

Salt is power: it makes things happen. It has started revolutions, created trading empires, helped build great cities, from Venice to Liverpool. The simple chemical's centrality as an ingredient and a tool, a means of adding value and life to food, can be seen in all languages; in English, words from salary and sausage to sauce and salacious all derive from the Indo-European word *sal*.

Salt gets used as a metaphor across the world. In French, *mettre son grain du sel* – to 'add one's grain of salt' – is to add originality, inspiration, a conversation-changing thought. Salt is a purifier: Japanese and English use 'a pinch of salt' to test a boast or untruth. Salt stands for community and social cohesion: in Spain and Poland people say that you need to eat a 'pile of salt' with someone to be true friends.

In Michelangelo's *The Last Supper*, Jesus's betrayer Judas Iscariot has spilt the salt cellar in front of him. He is breaking the social contract, bringing bad luck. Salt is unique, valuable, crucial. It deserves a chapter of its own in any history of humanity.

WHAT IT DOES

The compound – NaCl, a.k.a. sodium chloride – is both simple and fantastically complex. Some of its mechanisms are still not yet fully understood but the most basic effect emerges when it is eaten. It binds easily to food molecules and then when it enters our mouth it moves into saliva. The sodium chloride passes through the porous epithelium to directly stimulate the cells of nerve endings in the mouth, which turn on our taste buds. Salt is vital to human physiology – for our nerves, muscles and blood health. But we don't actually need much salt for our bodies to function, far less than the average 8.5g a day that Britons eat.

Neuroscience tells us that neurons in the brain control hormones that regulate our salt intake, sending us to seek salt or to stop eating it, and that these desires are dictated by genes. Mice without the gene consume three times as much salt as those with it and swiftly develop high blood pressure. In fact, we need only perhaps 0.5g of salt a day for our nerves and muscles to function properly, though official maximum recommendations are twelve times that.

Only sodium chloride (forty per cent sodium, a metal, and sixty per cent chloride, an acid) actually tastes of what we call salt, but sodium partners with many acids to create other compounds chemists also call salts. Soap is a sodium salt. Amino acids, such as aspartate and glutamate, become salts to give cheeses and cured meats their must-have-more flavours. Hence monosodium glutamate and *umami*, naturally occurring in cheese or preserved meat, seaweed and tomatoes, cheaply turns bland flavours to exciting ones. Why do we grate Parmesan on pasta? To make a near-tasteless carbohydrate zing.

Our brain tells us to seek salt, as it does sugar and fats. It then rewards us through the hormone system when our taste buds inform that the quest has been successful. This basic mechanism is the cause of many ills, from poor health to human conflict. The Romans fought wars over salt works and supplies of salt; so did the Americans in their civil war. Taxes on salt have caused rebellions across the world, not least in the 1920s in British-ruled India. But salt is, of course, also the origin of some fundamental joys in food.

'Salt is the policeman of taste: it keeps the various flavours of a dish in order and restrains the stronger from tyrannizing over the weaker.'

Food historian Margaret Visser, *Much Depends on Dinner* (Grove, 2010)

30

In cooking, salt performs magical transformations. It will firm up floppy fish or meat and turn a watery vegetable into a crisp one. Salt lowers the freezing temperature of water, which is why it is crucial in ice-cream making – and useful for making an icy path safe. Like sugar, salt will blur bitter tastes and enhance others: most recipes for cakes or sweet biscuits include a little salt to add something to the

Salting away sour

Salt's ability to mask or cancel bitterness is just as dramatic as its taste-enhancing effects – and not yet scientifically explained. You can experiment with a mug of water. Squeeze a lemon in and it should be bitter enough to make you screw up your face. But half a teaspoon of salt stirred in will neutralise it.

flavour that we find hard to describe – depth? complexity? pizzazz? – though we know it's missing when it is not there.

Bread simply does not work, taste-wise, without salt – though 1kg of flour will need very little, just a couple of teaspoons of salt, a 1.2 per cent mix. Fresh tomatoes sliced with olive oil are magically transformed with a sprinkle of rock salt, as though the very spectrum of their taste has been widened.

Chef Samin Nosrat, Netflix star and author of *Salt Fat Acid Heat*, told *The Food Programme*, 'Using the right amount of salt – in most cases more than my health-conscious mother uses – is crucial to getting food to taste its most possible best. To be the apotheosis of itself. When it's salted properly you just taste it more. That was the most fundamental lesson I've learned as a cook and it has guided my career.'

SALT AND POWER

Salt, like sugar, has another property beyond what it does to our pleasure centres – an asset just as important in how it has affected human food culture and the economics of food trading. Without salt, no *jamon serrano*, no salt cod, no sun-dried tomatoes, no ham, no smoked fish.

Salmonella cannot grow in salt concentrations as low as three per cent, while the bacteria responsible for botulism dies at around five and a half per cent. So, fish is cured in a solution of around six per cent. That turns one of the most short-lived of foodstuffs into a protein package that will last for months. Salt turns a fisherman's or a meat farmer's work into something bankable.

The key to this is the basic, natural process of osmosis, whereby a permeable membrane will allow a substance to pass through until a balance is reached on both sides. It is how plants absorb water through their roots. Salt will draw water through until both inside and out are as damp as each other. The salting process also kills bacteria by drawing water out of the organism's cell until it dies and it dries the muscle tissue of fish or meat.

It is salt's qualities as a curing and anti-bacterial agent that have made it a geopolitical commodity, a player in the making of civilisations. Until canning and refrigeration, preserving fresh meat in this way kept it useful. Rather than having to eat it within a few days of slaughter, it became a portable, storable and valuable commodity. Even smoking, which also cures meats, is dependent on giving the flesh a salty bath before the process begins.

Being able to preserve meat means power and control. People on the move – traders, armies and navies – can be fed without the need to forage or loot food. Thus, salt became a means to power and a token of it; the word 'salary' comes from people, from Roman legionaries onward, being paid in salt. Immense wealth flowed from possession of the sources of salt and by the early Middle Ages, salt was as valuable as gold. Coastal cities thrived from being close to

'A good way to powder or barrell beef' from
***The Good Huswives Handmaide*, Thomas Dawson (1594)**

Take the beefe and lay it in mere sawce [a marinade of beer, salt
and vinegar] a day & a night. Then take out the beef and lay it
upon a hirdle [rack], and cover it close with a sheete, and let the
hurdle be laid upon a peverell or cover to save the mere sauce
that commeth from it: then seeth the brine, and lay in your beefe
again, see the brine be colde so let it lye two days and one night:
then take it out, & lay it againe on a hurdle two or three days.
Then wype it everie peece with linned cloth, dry them and couch
it with salt, a laying of beefe and another of salt: and ye must lay
a stick crosse each way, so the brine may run from the salt.

salt production areas, and then grew rich from trading in it and
foods made with it. Many communities and economies arose from
the salting of flesh to store and sell. But none can match the story
of salted cod.

BACALAO, CODFISH, SALTFISH, KLIPPFISK

Picture a twisted length of yellow, dry flesh, spiky and corrugated,
a bit like a piece of old mooring rope. Get closer and there's a
smell, a little ammoniac, not entirely off-putting. You can find
dried salt cod chunks in the back of a Caribbean grocery store in
any European city, stacked under a shelf or as whole split tails and
fillets hanging on string from a rafter. Touch it, and the muscle is

fibrous, bone-dry, tough, curiously light but 'as hard as a cricket bat', as they say.

Once the protein of the poor across Europe and of both free and enslaved workers in the Caribbean, as is so often in history, a food of necessity became a beloved staple. Atlantic cod is hard to find because of centuries of over fishing and most saltfish cooking in West Africa and the Caribbean now uses pollock and other more plentiful white fish. Quality traditional salt-cod fillets can sell for the price of the best cuts of beef – €25 a kilo in Spain, more in Britain.

Tradition has made salt cod a delicacy around the Mediterranean. It is a Christmas dish enjoyed in Portugal, Italy and Croatia, an Eid-al-Fitr celebration feast in North African countries. It remains the foundation of treasured snacks and culture-defining dishes in Nigeria, Ghana, Brazil and Angola, as well as the Caribbean.

Cod are migratory. Human beings learnt more than a millennium ago to turn the seasonal arrival of great shoals into year-round supplies of protein for their communities and others with whom they traded. The first method, still used, was simply to gut and split the fish and leave it to dry, killing the bacteria, on rocks or on racks in stone houses built to funnel wind. The remains of these can be seen all over Scandinavian countries and on some Scottish islands.

But this could take months. About 1,000 years ago, fishermen from further south introduced a new, speedier technology: salt drying. It emerged in Mediterranean countries where the sun was strong and salt was easy to manufacture. So salt was traded north to where fishermen found cod, in the north-west Atlantic, and then to Iceland and eventually the great banks off Canada. They salted the fish and traded them around the Europe-centred world. The fish and the salt made fortunes and built city-states.

SALT ADVENTURES

The historian Lizzie Collingham, in her book *The Hungry Empire*, makes a case for salt being key to the early growth of Britain as a trading nation. By 1615, 250 ships from the West Country were sailing every spring for the rich cod grounds of Newfoundland. Thousands of jobs were created in England in shipbuilding and rope and sail making. London investors got excited by the business, which could return as much as fourteen per cent per voyage. Merchant ships were commissioned specifically for the trade.

In 1664, James Yonge and his ship the *Robert Bonaventure* made a typical voyage. He sailed first from Plymouth 6,000 miles south, to Boa Vista, an island off Cape Verde on the west coast of Africa. In 1620, the Portuguese had introduced a population of enslaved Africans to the little-inhabited island expressly to make salt for the Atlantic trade. With its holds full, Yonge's ship made its way from Africa north-west to Newfoundland, in what is now Canada. He sold the salt to other ships and the *Robert Bonaventure* did some fishing of its own. With a full cargo of dried cod, Yonge next headed for the Mediterranean and Genoa in northern Italy, where the salt cod was sold. He probably loaded his ship with wine, currants and olive oil for the trip back to Devon.

Salted cod and meat were the fuel for the men of the ships, as well as their trade; they made months-long voyages feasible. By the end of the seventeenth century, Britain was a major sea power and its battleships and merchants carried barrels of salt pork, butter and beef – known in the Royal Navy as 'Irish horse' – to feed the crews. Very little changed until frozen and tinned meat became available in the nineteenth century.

Salting enabled other industries, too. As European nations started to ship enslaved Africans to work on sugar, cotton and tobacco plantations (and indeed on salt pans) in the Americas, it quickly became clear that it was more cost effective to import protein to feed this workforce than use their time and labour to produce their own food.

By the mid-eighteenth century, an enslaved person in the Caribbean was a valuable asset, worth, in twenty-first-century terms, the price of a family car. The enslaving industry started shipping in lower-quality salt cod – the 'red' fish that had a fungal bloom and the cod heads – to feed their African slaves. With cod now in demand throughout Europe, it was cheaper to send barrels of salted herrings to the plantations, supplemented by salted beef and animal heads.

A boom began in Ireland and the west of Scotland to supply the slavery industry: Britain alone had 800,000 enslaved people working on Caribbean plantations by 1800, half of them in Jamaica. In 1798, a parliamentary enquiry found that 84,782 barrels of herring – 90 million fish – were shipped from Britain to the West Indies every year.

Today in Jamaica, saltfish and ackee – a fried mash of that fruit, chilli and rinsed fish – is the national dish. Salt cod head is still used. Jamaican children are warned not to look at the twisted dried head of the fish before cooking – so ugly it will give them nightmares.

EATING SALTFISH

Salt doesn't just preserve fish: it improves and adds value to many, from anchovy and sardine to cod. Eaten fresh or from a freezer, cod is fairly bland, prized more for its firm and flaky texture and snowy

The hugely popular *Mrs Beeton's Every Day Cookery* has several recipes for fresh cod, mainly with French sauces, but only one for when it has been salt-cured. She notes that it needs soaking overnight and then should be served cold – presumably as a punishment – on the Christian day of penitence, Ash Wednesday, with egg sauce and parsnips.

colour than for taste: few people would buy a cod supper from a chippy and leave without vinegar, salt or ketchup. European cuisine has very few classic recipes for fresh cod but Portugal alone has over one thousand for salt cod. *Bacalhau* remains an iconic ingredient there as well as in the Caribbean.

The only place around the Atlantic where salt cod has never been a part of *haute cuisine* is Britain, perhaps because it was associated with the impoverished underclass. Shakespeare knew it as 'poor John'. But the salting and then rinsing of the fish is a wonderful thing: it brings out complex, intriguing flavour, as well as an extraordinary chewy texture. England's best-known seafood chef, Rick Stein, suggests serving 'dull' fresh cod only with strong boosters: a rioja wine sauce, or as part of a pork and mussel chowder. He would rather eat salt cod.

THE SALTWORKS

For millennia, salt was produced in just two ways. People could evaporate seawater with sunshine or fire, or mine the deposits left underground by ancient seas. The whiter a salt was, the more valuable – people knew that the salt-making process was dirty, especially in river estuaries, and worried about what was among the crystals.

'This whole trend for special salt is a kind of ironic reversal of history,' Mark Kurlansky, author of the bestselling book *Salt*, told *The Food Programme*. 'Because until the twentieth century, there was a huge variation in salt – they had trouble with consistency, with getting rid of impurities. Old recipes will often specify a

Salt your own cod

Saltfish is available in Caribbean supermarkets and some mainstream ones in Britain. But you can salt a piece of fresh cod – or any white, non-oily fish – yourself, though you'll find it hard to get the eighty per cent reduction in water and weight that traditional salt-drying achieves.

Dry a thick fillet of fish with kitchen towel, place it in a sealable box and cover it all over with a thick layer of salt for 24–48 hours. Drain the liquid from the box every now and then. At the end of this process you can use the fish or place it on a rack in your fridge to let it dry more. Hang it from the ceiling – or, less authentic but more practical, store in the fridge or freezer.

To cook it, it must be rigorously desalted – some recipes suggest soaking in fresh water for 24 hours or boiling it twice and throwing the water away. Countless recipes then await you – cod *brandade*, a saltfish pate or classic Jamaican saltfish fritters with a chilli sauce are good ways to start.

certain kind of salt because there were so many salts that were completely different.' He points out that people used to pay more for salt the whiter it was, whereas today a dirty grey salt like that from the salt marshes of Guérande in Brittany is among the most expensive of all.

Salt becomes whiter with rinsing: the industrial process is no more complex than that. The only real improvement to salt came with the modern age, when in 1911 the Chicago businessman Joy Morton added an 'anti-clumping agent', magnesium carbonate, eliminating the age-old problem of having to keep salt completely dry. Morton's salt with its slogan 'When it rains – it pours' was a predictable success. For decades, recipes in the US would specify using it and the company still thrives today.

TOO MUCH, TOO LITTLE

Getting our salt intake right is important. Two million people die each year around the world because of too much salt in their food. Fear of the chemical is clearly justified and it has made money for the food industry. 'Low sodium salt' mixes potassium chloride with sodium chloride. Potassium has the opposite effect of the sodium – it lowers blood pressure. But eighty per cent of the salt we eat comes in the form of additions to processed food. Promises of 'low in salt' on packets have been shown to have little effect on overall consumption: the food writer Bee Wilson says these are often just marketing slogans, for all the value they have in influencing our diets.

Cooks can try using soy sauce instead of salt (it contains salt, but less of it). Using the flavour enhancer monosodium glutamate, feared and loathed by Western consumers but ubiquitous in East Asia, also makes salting less necessary.

Iodised salt has had potassium iodide added to counter a common deficiency that affects the thyroid's function. Since the 1930s, iodine-enhanced salt has been common: it was one of the first mass health interventions through diet and in many countries iodising salt is still mandatory. In Germany, where it is voluntary, a 2016 survey found that a third of adults and children suffered mild to moderate iodine deficiency. The first sign of thyroid problems caused by this is a swelling in the neck – it used to be called a goitre. Weight gain, hair loss, heart arrhythmia and brain dysfunction can follow.

40

Cut your cooking salt?

Studies show that it's a bad idea to leave salt out of your cooking and let people decide how much to add. People who add salt at the table tend to consume more than if salt had been used in the same dish while cooking it. 'Most people criminally under-salt,' especially in cooking water, says the chef Samin Nosrat, author of *Salt Fat Acid Heat.* 'It's the most important flavour enhancer we have.' But that doesn't mean use more of it, she says. 'Use salt better. That means add salt earlier and use it throughout your cooking. If you don't salt your meat far enough in advance to give it time to distribute, it won't season it properly. It needs time to get in. In Japan, salt is built into the dish, or other ingredients. Salt is not on the table, it's in miso or soy sauce.'

KILLER SALT

By the mid-twentieth century, salt itself came to be seen as a health problem – the poison that, in excess, it clearly is. Cardiologist Professor Graham MacGregor says: 'The amount we eat in the UK is a chronic long-term toxin, slowly putting up our blood pressure over many years. And raised blood pressure is the biggest cause of strokes, heart attacks and blood pressure: salt is playing a major part in all those people dying.' Today, most of our salt consumption comes from processed foods, ready meals and food eaten outside the home: research in the United States and UK has put the figure at seventy per cent of all adult salt intake.*

Professor MacGregor has spent his career campaigning for reductions in salt and sugar intake. He deplores the rise of salt use in gourmet products like chocolate and mocks the claims of designer salt makers. 'There's no difference in designer salts – they're all between ninety-six and ninety-eight per cent sodium chloride. Some are labelled as organic but it's nonsense to call a chemical organic. The water is boiled. The trace elements are in such small amounts they can't make a difference.'

Whether such salts come as flakes or as powder is, equally, not a sign of any particularly artisanal method of making the salt – just a guess at how the consumer would like it to appear. Flaked salt is of course bulkier, so packets can look like better value. Its crunch is attractive, but any claim that it 'dissolves in a different way on the

---

* 'The Bulk of US Salt Intake Comes from Processed Foods', American Heart Association research quoted on cardiosmart.org (2018). Original research unavailable. UK research from 2008/9: www.ncbi.nlm.nih.gov/pmc/articles/PMC3561609/

tongue' is just marketing talk. One advantage of flaky salt is that it doesn't clump up like fine salt, so anti-caking chemicals do not need to be used.

Most chefs back up MacGregor's view that designer salts are not practically different; they're for the service, not the kitchen. Mark Hix doesn't use them in his normal cooking, he uses table salt. 'It would be a waste to throw Halen Môn into the potato boiling water,' he laughs. He likes the brand's vanilla-flavoured salt – it works well with seafood and duck – but admits you could get the same effect with a vanilla pod.

Companies like Halen Môn in Anglesey pump in raw seawater and heat it: there is nothing different about the essence of the process and the shape of crystals is about how it is dried, not how it was originally obtained. The best that can be said about the artisanal salt makers, it seems, lies in their claims to be more energy efficient and more gentle on the environment than industrial saltmakers. Cornish Sea Salt told *The Food Programme* that when it pumps seawater into its factory it only takes a proportion of the salt out, returning the rest to the ocean in a condition that won't damage sea life.

The presence of invisible microplastics is another challenge for salt producers. Ninety per cent of seawater-derived salt contains plastics, a 2018 survey found, while another states that the average consumer eats 2,000 microscopic pieces of plastic in salt each year. Halen Môn says that its multiple filters will remove microplastic particles down to those bigger than ten microns, a seventh of the width of a human hair.

WHAT A MARK-UP

Industrially mined rock salt for winter roads sells for as little as £25 a ton, *fleur du sel* from an artisan maker for £75,000. They are exactly the same chemical and so fraudsters have, inevitably, been attracted. So lucrative is the artisan salt business that some corner-cutting, not-strictly-illegal-but-certainly-dubious practices are now fairly commonplace. There is little regulation or inspection of such businesses and it seems that adding industrially produced salt – often imported from abroad – is not unusual, just as using mass-produced grain alcohol is normal among many artisan gin 'distillers'.

Consumers won't always put up with these scams. In 2017, the successful company Hebridean Sea Salt, suppliers of Sainsbury's and others, was forced to close after it was revealed that its product, made 'by hand' on the Isle of Lewis, was made from eighty per cent industrial, imported salt.

All the same, premium salt – whether transported in a bamboo stick from Myanmar or pink and mined in the Himalayas – is one of the twenty-first century's most enduring food obsessions.

The most expensive of all

Amabito no Moshio is made beside the Seto sea in Japan using an ancient method involving infusing seaweed with concentrated seawater. After evaporation and roasting, the salt comes out tangy and caramel-coloured. In the UK it sells for £4.95 for a tiny packet – 50p a teaspoonful.

Halen Môn, Britain's most successful brand and winner of multiple international awards, sells most of its product to countries on the Mediterranean – from where much of our salt used to come.

One thing is clear: salt is not going away; we need it and we like it, to the extent that surveys show we all under-report how much we add to our food. Professional chefs – if you question them – confess

Should I salt the water?

Most people believe that we add salt to water to make it boil more quickly and increase the boiling temperature. It will do that but only by a degree or two Celsius – not enough to make much difference to most food. Salt won't boil an egg – which starts to harden at 65°C – any more quickly but it does provide a safety net. It makes the water denser than the liquid inside the egg. If you've cracked the egg the white is less likely to escape and it will coagulate more quickly if it is exposed to salty water. But you need a tablespoon or more, not just a pinch, in the pan for that effect.

Similarly, to get a litre of water to boil two degrees above 100°C, you need to add 230g of salt. With pasta, rice or boiling vegetables a pinch or two will have no effect other than to pre-salt the food – but it will do that more evenly than you can later at the table. We're told we should all eat only a teaspoon of salt a day; if you put a couple of teaspoons of salt in the pasta water, the pasta will only absorb a quarter of it, but around the table people will feel less need to salt the dish.

that they use far more salt in everyday cooking than they would admit to if publishing the dish's recipe. Necessary for life, necessary for taste, salt is for ever.

CHAPTER 3

FATS OF THE LAND

OIL AND FAT

'I thought fats were just fat, but in fact fat is probably
the most misunderstood food on the planet.'

Pat Whelan, whose beef dripping was awarded
Supreme Champion at the 2015 Great
Taste Awards, speaking on
The Food Programme

They are called butteries or rowies. The first one I was given came with the warning 'it's a heart attack on a plate!' In north-east Scotland, they're known as fishermen's food – handy energy packs for eating during long shifts out at sea. But the devotion to them goes much wider. A chewy, layered light pastry filled with animal fats, they are baked until just browning and then eaten. Best hot and perhaps spread with honey, a bite into the buttery is a chew not a crunch, and the salty butter in it is the dominant taste. Coronary threat or not, this slab of just-cooked oily dough satisfies a deep and primal need. Afterwards my lips are greasy: I ask for another. 'I never, ever make enough,' grins the baker, a man descended from generations of Moray Firth trawlermen. (See the end of this chapter for a recipe for butteries.)

Butteries are not unlike another layered pastry, more sophisticated but just as beloved: the croissant. Most recipes for a Scottish buttery are about forty per cent fats, and so is a proper traditional croissant, though some recipes go higher. The French layer-roll uses butter; the Scottish one butter and lard. But the important point is that these are foods that flaunt their fattiness – and without it they just would not work.

In the twenty-first century, most foods made with or containing fat – which means most foods – are busy trying to hide this embarrassing ingredient. Fat has become a bad thing, in the tick-box approach to our modern diet. Yet so much that we love to eat – from crisps to chips, beef steak, fried chicken, chocolate and cakes – is dependent on the texture and 'mouth-feel' (as food tech calls it) brought to you by fats and oils – the latter being simply the liquid state of the first.

We need fats and oils – compounds of glycerol and fatty acids, to be technical – in our diets. Along with protein and carbohydrates, they are one of the three crucial macronutrients without which a human cannot survive. Some hunter-gatherers in extreme environments can survive on almost no carbs but they couldn't without fat. The Inuit communities who still live traditionally in polar regions fill the carbohydrate gap with animals and fish, while richer human beings on no-carb or 'keto' diets find the regime very hard. Fats are the fuels that repair and maintain our liver, our nervous system and particularly our brains. In Britain, official advice is that calories from fats should make up thirty-five per cent of our daily energy intake – we can eat up to seventy grams of fat a day, more than a quarter of a standard pack of butter.

FAT, FABULOUS FAT

Butchers and chefs know that when it comes to meat, the taste lies in the fat. When cooked, it melts into the meat, providing what we perceive as flavoursome juiciness: that's why beef for mince and steaks is sold at two fat grades higher to restaurant suppliers than it is to supermarkets, where most health-conscious customers still like to see a lean meat with only a few white veins of fat. These are known admiringly in the trade as 'marbling': the more the better.

At a session organised by the government agency Quality Meat Scotland (QMS) to inspire chefs and restaurateurs, I watched a chef cut two lean, fatless steaks from the backbone – one venison, one beef. He quickly fried them rare. We were given slices on unmarked plates and asked to say which was which: we could not. Without fat, the fillet meat was just meat. Laurent Vernet, then the QMS culinary expert, explained that flavours lie in the fat. Lamb raised on saltmarshes, a premium delicacy known for its subtle salty flavour, is indistinguishable from ordinary grass-fed lamb if the fat is removed.

Lard, the fat of pigs, is a gourmet substance in Southern Europe: quality Iberico ham may be served with a good trimming of snowy white fat to be eaten along with the red meat. But in Britain, the word is a comedy insult: unhealthy and obesity-engendering. In 1974, the average Briton ate 75g of lard each week; today it is only 5g – chiefly concealed in ready-made products. But people who love food know the value of lard, especially in pastry, and the more expensive your bacon the more fat there is on it.

Chef Jeremy Lee remembers his grandmother in Dundee frying beautiful little pancakes in lard – 'really crisp on the outside, with this gorgeous fluff on the inside, and tons of butter. Utterly delicious.' 'Did that taste piggy?' asked *The Food Programme*'s Oliver Thring. 'No! Lamb fat can go rancid quickly, beef fat becomes quite pokey with the age of the meat, but pork you eat very fresh: it's sweet, clean and delicious. It's a blank canvas.'

DRIPPING TRIUMPH

When Pat Whelan, a fifth-generation butcher in Clonmel, Ireland, asked his mother what the previous generation did with all the fat now discarded in the shop during the cutting up of a beef carcass, she told him to bring some of it to the kitchen. On her instructions, he selected different types of fat – the suet from around the animal's kidneys and hip bones, the back fat from around the best cuts of beef and the body fat.

'When we started to actually melt the fat,' Whelan told *The Food Programme*, 'we noticed that fat from different parts of the animal had different flavour.' And when the different fats were blended it produced an extraordinary taste experience: 'A belt of roast beef, of Yorkshire pudding and you get roast potatoes, almost in one bite. But there's nothing added.'

The dripping was not new, just the rediscovery of an artefact that was commonplace in kitchens and butcher's shops until sixty years ago. Pure vegetable oils were expensive and rare then, and most people only knew them in the form of margarine. Our grandparents' frying, roasting, pastry-making and cake-baking were all accomplished with animal fats: butter or the rendered fat of the animals we ate, from poultry to pigs and cattle. They benefited too from the omega-3 fatty acids in which grass-fed animals are rich.

The Whelans started selling their dripping through their shops in west Ireland and online. A few months later, the world gave its verdict. 'When we tasted it, the final judging panel, everyone was silent. It was a reverence and in a room full of really gobby food writers, that's quite something,' Felicity Cloake told *The Food Programme* of her experience of Pat Whelan's beef dripping. She was a judge in the 2015 Guild of Fine Foods' Great Taste Awards, where Whelan's new product managed to beat 10,000 others – from mueslis to ciders – to become that year's Supreme Champion.

Cloake now cooks with dripping and she confesses to liking some simply spread on toast. Pat Whelan is delighted. At the Great Taste Awards his prize was announced with the words: 'Fat is back.' His mother, seeing the acclaim given to a recipe made by the family for over a hundred years until fashions changed, burst into tears.

GREASE IS THE WORD

You can still find lard and dripping on the shelves of larger supermarkets but they occupy a tiny corner compared with the ranks of bottled corn, sunflower and other vegetable oils. Animal fats are not so different to use and they're generally cheaper than vegetable oil. Ghee, the standard animal oil in Indian cooking, is made by boiling butter to remove water along with the lactose and casein that may trouble people allergic to ordinary butter. It is similar to what the British call clarified butter and it is another useful cooking fat as its smoke point (where an oil breaks down and there is an increasing risk of fire) is a good 80°C higher than that of regular butter.

In a supermarket, I bought 250g blocks of each of the three traditional British cooking fats displayed side-by-side in the fridge by the butter: Stockwell & Co. Lard (39p), Britannia beef dripping (70p) and vegetable oil-derived Trex (95p). Back home, I melted them down and discovered they all had different flavours.

I par-boiled and sliced potatoes and then fried them in three different pans – one per fat – for my family. They'd normally have potatoes fried in sunflower oil. Everyone liked the chips and in a blind taste test, no one found detecting a difference easy. All three brands seemed to have done the job well – but the beef dripping-fried potatoes won when put to a vote.

The problem came when I told the table that two of the fats were derived from animals. That did not go down well at all. Why? I asked. 'Dirty,' said my wife. 'Cruel to animals,' said my daughter. (Neither of them is vegetarian.) But, I said, the animal was killed for its meat. This just stops waste. (Some animal fats go from the slaughterhouse into the making of soap and detergent, but much more goes to landfill.) My daughter vowed to start checking the label on the soap; my wife said that now she knew she was never going to eat chips unless she cooked them herself.

So why are the British so suspicious of fats and oils? One cause may be linguistic. In English and some other European languages, the word fat as in overweight and fat as in meat is the same. This, of course, can be taken to imply that fat in food becomes fat on your body, which is not true at all.

Fats and oils break down in the gut into fatty acids. Fat on our bodies is produced by the breakdown of carbohydrates, including sugar, as a way of storing surplus energy. It's notable that in Spain where fat in food is 'grasa' and a fat person is 'gordo', there is less fear about fat and oil in cooking. Pig fat is treasured; olive oil is central to all cuisine. The percentage of the population classed as overweight is fifteen per cent lower in Spain than it is in the UK.

Much prejudice against food fats seems to come from class issues and xenophobia. Foreign foods, and even foreign people, were 'oily' or 'greasy' to many Britons who grew up in the mid-twentieth century. Food habits abroad were viewed as unhealthy or even morally dubious. 'In the 1980s, when my father wished to express disapproval of a dish, the word he reached for most frequently was "oily",' writes food historian Bee Wilson. 'When our family bought a takeaway curry he would sometimes comment that the chicken biryani or the poppadoms were 'not too oily'. This was high praise. Oiliness to my father signified food made by a cook who could not be bothered to skim away the excess fat on the surface of a dish of Irish stew or to blot the underside of a fried egg before serving it.'

Such snobbish concerns were not just a British phenomenon. In 1968, the American food writer M.F.K. Fisher remembered her grandmother's 'Victorian neuroses' about food: she had a deep suspicion of salad – no more than roughage and a 'French idea' to boot – and of olive oil. In place of mayonnaise, she made something called 'boiled dressing', which was simply vinegar thickened with flour and a little salt. It was bad enough, wrote Fisher, to 'harm your soul'.

Bee Wilson goes on to write that the oiliness her father feared is now 'everywhere' – by stealth. In her book *The Way We Eat Now*, she interviews scientists looking at changes in what's known as the Global Standard Diet. The thing that surprises them most is the enormous rise in the consumption of vegetable oils over the last fifty years, with soybean oil at the top of the list. Because they are cheap, soy beans have added more to global calorie consumption than any other substance, far outpacing the increase from our ever-rising sugar habit. Soy oil has now become the seventh most eaten food in the world, closely followed by palm oil, without, as Wilson says, anyone ever really desiring them.

That is true of the richer world. But much of the rise in consumption of soy oil and other cheap vegetable oils is seen in

Fats and cholesterol – an explainer

Saturated fats. Solid at room temperature and obtained mainly from animals, coconut and palm oils. Found in meats, dairy, pastries, cakes and puddings. Linked to increases in blood cholesterol levels, which may be a problem in heart health.

Unsaturated fats. Usually liquid at room temperature. Sources include nuts and seeds; vegetable, olive and nut oils; soft vegetable spreads; oily fish and avocado. Poly-unsaturated fats (where the molecules are bonded differently) may be better than mono-unsaturated fats.

Trans fats. An industrial process of adding hydrogen into the fats makes them much more stable, but with a proven risk to heart health. Long-life pastries and biscuits may still contain them – look for phrases like 'partially-hydrogenated vegetable oils' on labels – but generally they have been banned or phased out.

Cholesterol and lipids. Fats circulate in our blood as cholesterol and triglycerides: both are called lipids. They have essential roles in the body, repairing organs, transporting energy and making hormones. However, in excess they are harmful. Cholesterol and triglycerides travel around the blood in lipoproteins. Low-density lipoprotein (LDL) carries most of the cholesterol in our body from the liver to the cells in our organs. High-density lipoprotein (HDL) takes excess cholesterol. HDL cholesterol is called good, LDL bad – they are believed to counteract each other.

parts of the world that were until recently among the poorest. Oil was always a luxury, sparingly used, and so it is a great pleasure, and a flaunting of affluence, to use it liberally. That's why modern Chinese fried foods are often seen as scarily oily by outsiders. But the truth is that Western vegetable oil consumption has risen hugely too, though our oil tends to be hidden inside the products that use it, not bubbling in the wok.

GOOD FAT, BAD FAT

Nutrition is complex, informed by an evolving multi-disciplinary science, with a worried audience and greedy industry both eager for simple solutions. We who worry want a simple 'good' or 'bad' on labels along with straightforward advice we can follow. With fat, as with cholesterol, gluten, sugar and all the other scare substances that have seized the public imagination in recent years, the truth is much more slippery than that.

The science around fats and cholesterol is often counterintuitive, sometimes contradictory. Even qualified dietitians can seem foxed by it. 'Healthy' slimming diets can *increase* LDL (low-density lipid or 'bad') cholesterol as semi-starvation can upset hormonal systems and adrenalin production.

Saturated fat may chiefly be a problem when you eat it with the wrong foods, like highly processed carbohydrates. It's also no surprise to find that people who eat lots of whole-fat yoghurt are generally healthier than those who eat the same fat in the same quantities but in the form of butter – yoghurt eaters tend to have healthier lifestyles, but the fat is the same. Saturated fats are an issue in type 2 diabetes but not in the wholly negative way once believed; there's evidence that low-carb, high-fat 'keto' diets can actually help sufferers from diabetes.

Excess fat in the diet has long been known to have a bearing on heart health. But, after decades of wagging a finger at all non-vegetable fats, science has now decisively changed its mind. The research that started in the 1950s, which claimed saturated fats led inevitably to higher cholesterol and heart disease, has been shown to be deeply flawed. What has been called the 'French paradox' by American scientists reflects the fact that in France – as in many Mediterranean countries – people may have high cholesterol levels and high consumption of saturated fats but low instances of heart disease. Much of the research ignored other factors, like smoking. In 2014, a review of seventy-two different studies of heart disease and saturated fats, conducted by the British Heart Foundation, concluded that no link existed.

But in spite of the review showing no link, government advice remains in place to limit the amount of fats eaten, especially saturated ones. Such advice has always been skewed by the fact that the food industry has had a hand in devising health tools for the public, such as the government-approved 'Eatwell Plate'. This gives a graphic representation of what comprises a good diet. Generally, the British public consumes about twice the amount of fat required for a healthy diet; but nowadays it is mostly from vegetable oils.

UP AND DOWN WITH THE COCONUT

Our recent, strange relationship with the coconut tells us a lot about the confusion that our intense interest in science and diet can entail. Coconut water and coconut oil were subject to a fast-onset health fad that led to a world shortage of coconuts in 2015; in Britain, sales of the oil increased sixteen-fold, while the world price of coconut water increased by more than twenty-five per cent each year during the second half of the decade.

Coconut oil was credited with everything from reducing human belly fat to staving off dementia. In these stories, the phrase 'has been linked to' as in 'coconut oil has been linked to reduced hair loss in men', ought to put consumers on red alert. It usually means a speculative scientific paper has been written; its hypothesis yet to be proved. Coconut oil, like acai berries and goji fruits, also apparently 'bolsters the immune system', a health claim popular among manufacturers and influencers because it is scientifically impossible to prove. As ever, a counter-narrative followed: in 2018, the *Guardian* newspaper ran a story quoting a Harvard professor calling coconut oil 'pure poison'.

Karin Michels, an epidemiologist, simply reminded people that coconut oil is, and has always been, eighty to ninety per cent saturated fat (about twice as much as lard from pig fat). Whether it is good or bad for you is a question whose answer seems to change with every scientific paper produced. And when 'superfood' plant oils retail from £10 a litre (coconut oil) to £300 a litre (argan oil, from a Moroccan plant: it 'may help slow the ageing process' but certainly tastes great in salad dressing), there is no lack of incentive for manufacturers to keep sponsoring scientific papers that provide new selling points. More important, perhaps, as the sceptical food writer Felicity Cloake puts it, 'Coconut oil is great if you want your food to taste like a Bounty bar.'

OLEOMARGARINE, COAL OIL, WHALE'S BLUBBER AND CHEAP SEEDS

One of the worst side effects of the fat phobia that consumed the rich world for so many years was the terrible things we did to our diets – and ourselves – in order not to eat 'unhealthy' fats. The chief aberration was called margarine. Anyone who grew up in mid-twentieth-century Britain will not forgive those who made them eat marge, a clammy paste made of cheap vegetable oils and animal fats. We could have been eating butter in our sandwiches, whose vitamins and fatty acids make it advisable for those whose bodies and brains are still developing. While eating less butter may still be a good idea for older people at risk of heart problems, it now appears that generations of young British and American children's taste buds need never have been assaulted with margarine.

In 1869, the French Emperor Napoleon III offered a prize for a cheap butter substitute for his army and the country's urban poor to eat. Oleomargarine, made from beef fat, was the answer. It became popular in the cities of the late-nineteenth-century United States where it was sold as more hygienic and 'modern' than butter, which, to be fair, was often adulterated by unscrupulous dairies. 'Marge' was not universal in Britain until the Second World War, when it replaced butter under rationing. By 1942, each adult was allowed 2oz (57g) of butter a week but 4oz of margarine and another four of 'cooking fat'. By then margarine was made from a mix of various vegetable, animal and chemically manufactured fats, including a wax derived from coal and an oil from the blubber of whales.

Adding yellow dye and promising largely spurious health benefits made margarine more popular than butter in the United

States and Britain (though never in France) by the late twentieth century. While the spreads could be lower in cholesterol than pure butter, the oils were often hydrogenated to raise their melting point, making them solid and stable. These trans fats are now known to be far more harmful than any fat in butter or lard. Modern margarine has been reformulated to exclude trans fats, which are illegal in some countries.

The ongoing sat-fat obsession led scientists and doctors to largely ignore what was really causing the massive rise in incidences of heart disease, cancer of the digestive tract and diabetes. There was also a great increase in the consumption of refined sugars and salt, mainly in processed foods, and the two issues were directly linked. Fat is taste and if you remove it you need to replace it with something else. Thus, low-fat products, from mayonnaise to biscuits, tend to be higher in sugar and salt than the regular versions.

But it was TFAs, or trans-fatty acids, that were perhaps doing even more harm. Trans fats came to dominate food manufacture, thanks to a simple process perfected at the end of the nineteenth century. By attaching hydrogen atoms to oil molecules in the high-temperature process called partial hydrogenation you could raise the melting point of all sorts of previously useless oils, making them more stable and suitable for manufacturing everything from soap to axle grease.

THE WAR ON ANIMAL FAT

In 1911, Procter & Gamble, then a soap and candle maker, spotted how hydrogenation could be applied to food oils. It made a vegetable fat, Crisco, from cottonseed oil, using unwanted seeds from cotton mills. Tinned Crisco was an immediate hit in US households, as it was cheaper than the pork lard it replaced and lasted two years at room temperature before it went off. A replacement for butter followed: margarine precisely hydrogenated to remain solid at room temperature but melt on warm toast or in the mouth. 'Better than butter' was an early slogan – and manufacturers trumpeted the cleanliness of the substitute: 'Buy Swift's Premium Oleomargarine for its Goodness. It is healthful, nutritious and delicious,' went one 1918 advert.

Swift's was then made from beef fat, but by the 1940s almost all margarine was made from cheaper vegetable oils. Hydrogenated, these usefully mimicked the properties of pork and beef fats, so they went into halal, kosher and vegetarian foods too. By the 1960s, sixty per cent of American vegetable oils used in food were partly hydrogenated. Meanwhile, lobbying and health advocacy campaigns successfully demonised animal fats over the following decades: papers discovered by University of California and other researchers recently showed that some of the scientific work behind these campaigns was paid for by processed food companies eager to exonerate trans fats and to divert blame for rising obesity levels in the US away from sugar.

In the baking business, the oils were crucial for the new fast-mixing and proving techniques that needed oils that kept their solidity at widely different temperatures. The fact that the fats didn't go off, like butter or lard, gave bread and pastries a longer shelf life. They also had higher smokepoints, making them safer for frying. Best of all, hydrogenated oils – whether they came from whales, cattle or palm trees, soya beans or rapeseed – were fabulously cheap.

As with so many additives, once the food industry had found the cheapest method – catering fats based on hydrogenated vegetable oil (HVO) are about twelve per cent the price of butter – it set about justifying it in other ways. HVOs were sold as 'vegetable fats' or 'shortening', which certainly sounded nicer than animal fats. And it worked. British home cooks turned from lard and butter to margarine and vegetable fat.

It took decades for these 'clever, hygienic, modern' products to be exposed for what they were. The industry was fiercely resistant when journalists, including those on *The Food Programme* in the early 1990s, heard the dissident food scientists and started ringing alarm bells.

In 2006, the tide finally changed. Newspapers woke up to the message the researchers had been trying to deliver. Hysterical headlines talked of the 'killer fat', the 'Franken-fat that will not die' that was 'furring up our bodies like old kettles' and linked trans fats to disorders from Alzheimer's to autism. One researcher produced a big-brand packaged muffin, made with HVOs, which he said he'd bought ten years earlier. Out of its packet, it seemed as 'fresh' as it had ever been.

It is now generally agreed that trans fats, like saturated fats, *raise* cholesterol levels, put 'plaque' on our artery walls and thus in some cases bring about heart attacks. Our bodies find the hydrogen-altered oils hard to break down – in a standard campaigner's formulation: 'Would you melt Tupperware and put it on your toast?' Or, as the Food Standards Agency put it, in language agreed with the food manufacturers when official advice was changed back in the early 2000s: 'The trans fats found in food containing hydrogenated vegetable oil are harmful and have no known nutritional benefits.

They raise the type of cholesterol in the blood that increases the risk of coronary heart disease. Some evidence suggests that the effects of these trans fats may be worse than saturated fats.'

PALM OIL AND OTHER SINS

The packet of Trex I bought in Tesco says proudly 'no hydrogenated vegetable oils'. The label says this cooking fat is made from palm and rapeseed oil (and it admits it does contain some trans fat, 'like all vegetable oil'). Palm oil, which is one of the world's most popular cheap vegetable oils, has for years been criticised on sustainability and environmental grounds: grown in the tropics, mainly Malaysia and Indonesia, millions of acres of rainforest, crucial animal habitats and subsistence farming communities have been destroyed for the big palm plantations.

63

'It's very, very hard to boycott a product that's ubiquitous … from shampoos and soaps to food.'

Professor Chris Chuck, chemical engineer at Bath University, speaking to *The Food Programme* in 2019

In Britain, palm oil is ubiquitous: you can find it in everything from a Cadbury's Creme Egg to almost every biscuit and a whole host of other processed food and ready meals aimed at vegans and 'healthy' eaters – 'free from' products, as the industry calls them. Many of these products once used butter but the rise of fears around lactose – the sugar in milk – and animal fats encouraged the substitution.

Industry likes palm oil because it is cheap and stable – 'flavourless, odourless, Vaseline-like', as Sheila Dillion puts it. Fifty per cent of everything in our supermarkets contains palm oil, from toothpaste to cleaning products and cosmetics, as well as food. Concern from customers worried by reports about habitat loss and the plight of orangutans is addressed in the usual half-honest corporate way: the law says that palm oil must be listed, so labels use a host of euphemistic words to do so. Vegetable oil, diglyceride, PKO (palm kernel oil), palmate, stearate, stearic acid are just a few of them.

Many mainstream brands gloss their palm oil use, even if they are not shy about talking about it on the packet, by certifying their products with the RSPO label. Set up by the food industry and interested governments in 2004, the Roundtable on Sustainable Palm Oil now certifies ninety per cent of production. But it has many critics. Greenpeace, for one, calls the organisation 'about as much use as a chocolate teapot', pointing out that it failed to stop destruction of forest for new oil palm plantations until 2018. In response, the RSPO claimed it has 'one of the world's strictest sets of criteria regarding deforestation of any commodity standards' and that it has 'the best system available' to halt deforestation and destruction of important conservation areas.

The devastating fires that have swept Malaysia and Indonesia in recent years are in part blamed on the palm oil plantations. The fact that, in 2021, forty per cent of palm oil production is by smallholders, often impoverished, does not make teaching better practice easy. Many of the more aware eco-labels – like Divine Chocolate, who we look at in the chapter on cocoa – boycott palm oil, RSPO-certified or not. But this is not always a simple answer – under pressure, manufacturers may well switch to another vegetable oil, whose farming could potentially be equally ecologically damaging, requiring even more land for an equivalent yield.

Katie Major discussed this on *The Food Programme* in 2019. She is a campaigner and conservation psychologist at Bristol Zoo and has visited the Malaysian plantations: 'It's an incredible thing to see, it feels like these plantations never seem to end. Just rows and rows of these African oil palms.' But she does not boycott palm oil: she supports sustainable production. Oil palms produce four to six times more oil per hectare than other plants, she says: switching to other plants could be more damaging.

The solution may lie in manufactured or synthetic oils. These are derived from seaweed, algae and agricultural and food wastes, breaking them down with yeast. Researchers at the University of Munich say they have replicated all palm oil's useful attributes: all that is needed now is backing and government action against climate-damaging nut and seed oils to help 'yeast oil' find its place in the market. But this substitute doesn't come without its own problems: the yeast must be fed with sugar, which in turn has a climate cost in its production.

ANIMAL, VEGETABLE OR MINERAL

The animal fats vs. vegetable oil argument continues. Today, the animal lobby appears to be winning, even in the face of the current trend towards vegetarianism. There may be fewer greenhouse gas costs in animal fat production than in imported vegetable oils, if the animals are raised on locally produced feed. Of all the plants used for vegetable oils, only rape ('canola' in the States) is manufactured on any scale here in the UK. Health worries still emerge. The eminent diet doctor and author Michael Mosley recently became concerned about what happens when you heat various cooking oils, so he conducted an experiment in the laboratory raising the temperature of goose fat, butter, olive oil, sunflower oil and corn oil.

Cooks tend to use the last two most often for frying, because the temperature at which they smoke and break down is much higher. Fast food chains all fry with corn and other vegetable oils, as do most English chip shops. Further north in Britain, beef dripping is often still used because customers prefer the taste.

'I had fondly imagined that if you're cooking with sunflower oil it is going to be better for you, but that is absolutely not what the research showed,' Mosley told *The Food Programme*. 'Sunflower oil and corn oil are quite unstable at high temperatures and they produce aldehydes.' These are known toxic substances, implicated in everything from allergic hypersensitivity and respiratory problems to cancer and liver disease.

'It turns out the foods we've traditionally regarded as the saturated fats [like lard, butter, beef dripping and goose fat], tend to be much more stable when you're frying with them, and we found very few aldehydes in them.' Mosley says that he has now reverted to using olive oil for frying at home. 'I also happily use goose fat and [pig] lard because lard is actually very rich in monounsaturated fat.'

Mosley says that the food industry has largely ignored his findings, though medical practitioners are supportive. The most disturbing issue he found was that the vegetable oils that release aldehydes under heat tend to do so much more when reheated for a second time. In many chip shops and fast food outlets, the same oil is reheated again and again. It may be filtered but often only when it is visibly dirty is it changed. A Food Standards Authority report in 2017 showed that in some outlets the same oil is used, in the worst cases, for months.

OMEGA 3 AND OMEGA 6

The problems with our vast consumption of vegetable oil do not stop there. Another less understood issue arose with the replacement of traditional animal fats in our diet with vegetable oils. Most British people will make a face if you suggest they might enjoy such things as pork lard and beef dripping, and not just because they were so long erroneously associated with heart disease.

Some of the fatty acids derived from vegetables, animals and fish, like omega 3 and 6, are essential to the work of our complex organs. Deprived of omega 3, human beings start malfunctioning quite badly. Symptoms include poor sleep, lack of attention, cognitive disruption and a tendency to anger. (There's also dry skin, rashes and brittle nails, due to the oil's role in building cell walls.) Studies in prisons and mental health units have shown that people with a history of violence can be usefully treated with fatty acid supplements. Lack of omega 3 has been noted in patients with schizophrenia. Omega 6 is also important for cell health but deficiency of it is very unusual.

All oils carry the essential fatty acids omega 3 and omega 6 but vegetable oils have far more of the latter. Professor John Stein from the University of Oxford believes this is a huge problem. 'Ideally, we need one-to-one omega 6 to omega 3. That's probably what we had during our evolution. Nowadays, because of the mass shift to oils such as peanut, soybean, corn, sunflower, we have a great excess of omega 6 in our systems, which excludes omega 3.'

The fact that we have altered the diets of the animals we eat also affects us. Beef that has been fed traditionally, on grass, is an excellent source of omega 3 but the shift to prepared foodstuffs has pushed out the omega 3 and replaced it with omega 6. Chickens, once omnivores pecking round the yard, are now largely fed on corn, soy and other vegetable stuffs, thus losing out on the omega 3 they once got from worms and bugs. So worried is the animal industry about this that research is now going on into genetic modification of pigs – using DNA from a worm – to enable them to produce more omega 3 on a plant-based diet.

Professor Stein says that excessive intake of omega 6 will lead to it taking the place in the membrane of our brain neurons that should be occupied by omega 3. 'The problem is omega 6s don't have the same physical, chemical and electrical properties as the omega 3s: your membranes simply don't work as fast.'

It's this effect on the brain that Professor Stein thinks may account for all sorts of developmental issues in children, including slow reading and dyslexia. There is also evidence of a link to diseases associated chiefly with richer societies, like depression and senile dementias. Stein is convinced extra omega 3 – which can be obtained from fish (particularly the oily kinds such as sardines and mackerel), grass-fed red meat, seaweed and leafy green veg – can help stave off Alzheimer's.

His research confirms thinking that the pioneering scientist of brain chemistry, Professor Michael Crawford, first put forward in the 1980s and 1990s, when *The Food Programme* was one of the few places his research could get a hearing. He warned that the imbalance of omegas 3 and 6 was implicated not just in heart disease, but also in the rising epidemic of depression and mental health problems. Most scientists then were dismissive, but research since has proved him right: his *Nutrition and Mental Health: A handbook* (2008) is essential further reading.

Professor Stein's view of our future, should we fail to address the imbalance, is apocalyptic: 'I think the lack of omega 3s in our diet is going to change the human brain in a way that is as serious as [the effects of] climate change threaten to be.'

OIL OF THE OLIVE

Ninety-nine per cent of the world's olive trees are in the Mediterranean. They have shaped the landscapes of these countries and their history and culture. Olives and olive oil appear in the Bible, the Koran and Homer's epics, and the olive has symbolic use in many cultures and religions; the archaeology of the olive oil trade is rich with shipwrecks, pottery containers and olive milling tools and stones from Lebanon to Portugal and as far north as Scotland.

An entire hill of Rome, Monte Testaccio, is a landfill of smashed olive oil amphorae, some dating back more than 2,000 years. Most of them are are of a standard seventy-litre size, shipped from Spain. Others are from North Africa. In ancient Rome, as today, an awful lot of 'Italian' olive oil started on a tree in another country.

For millennia, olive oil and notions of civilisation have gone hand in hand. The art of writing appears in many Mediterranean cultures around the time that organised olive cultivation does – about 5,000 years ago. The miracle of the velvety golden-green liquid that can be extracted from those hard and unlovely fruits brought more than a way of living and eating better: it provided a currency, the basis of economies, rituals and trading empires. Not to have – or love – olive oil was a sign of a failure to join the most developed known society of the time, the Graeco-Roman one of the Mediterranean. Little has changed today: we cook with olive oil, sometimes unnecessarily, because fashion dictates it and admired chefs recommend it. To this day, it is the aspirational oil.

Roman legionaries wrote of the grim life on the Empire's northern frontiers, eating lard because the barbarians had no olives. (Nevertheless, shards of some of those seventy-litre amphorae have been found in the Roman forts on Hadrian's Wall, the northernmost outpost of all.) Even today, when older British food lovers talk of the dark ages in which they grew up, they will often point to the lack of decent olive oil. In the 1950s and early 60s, olive oil was only available in the chemist's, on the shelf beside the syrup of figs and extract of malt, though not usually intended for internal use. It was generally for clearing out your inner ear or conditioning hair. In 1861, Mrs Beeton,

70

How to kill a garlic lover

Many wonderful things can be done with olive oil, with its unique taste and its way of interacting with water-soluble substances like eggs and vinegar. But it can also be lethal. We've all seen garlic-, herb-, truffle- and chilli-flavoured olive oils: some of us have been tempted to make our own. What could be simpler than stuffing some garlic cloves in a pretty jar with extra-virgin? But you are dicing with *clostridium botulinum*, a lethal bacteria, which is found in soil, water and air. Oil's oxygen-free, environment is perfect for growth of this anaerobic bacteria, especially at room temperature. Garlic, with its high water content, is especially dangerous. Commercial olive oil with vegetable matter in the bottle uses acid or pasteurisation to kill the bacteria: but these processes are not easy to get right at home. Tomatoes are safe, because of their high acid content, and fully dried herbs may be, but keep the oil in the fridge. Don't get botulism!

dismissing olive oil as an obsession of 'continentals', wrote that it might be helpful against flatulence.

Today, most middle-class British households will have at least one bottle of olive oil and perhaps two: some extra-virgin for dressings and some cheaper type for cooking. Since 2005, we have consumed more oil from olives than any other plant; our annual consumption has gone up more than tenfold since 1991. The only problem is – and this story is as old as olive oil itself – it may well not be what it says it is.

SCAMMING THE OIL-LOVERS

This book is full of stories of food frauds, but the long history of olive oil contains more scams, thefts, poisonings, heists and sharp practice than that of any other food. The Sumerians, a civilisation centred in modern Iraq 5,000 years ago, had a royally appointed food standards squad, tasked with stopping olive oil fraud. But it is too easy and lucrative to substitute olive with other oils – and consumers are as gullible now as ever. Today, food labelling police are still busy. The authorities still have not worked out how to beat the cheats. In fact, new testing technology, like gas chromatography, serves merely to expose the extent of the fraud.

In 2012, a University of California laboratory set up to serve the olive oil industry in the state found from its tests that sixty-nine per cent of imported extra-virgin olive oil was not what it said it was (usually a lower grade of oil). Ten per cent of the Californian brands were guilty. It is not surprising: as one EU official told the author Tom Mueller, an American living in the oil-producing region of Liguria in Italy, the profits from passing off other oils as olive, or low-grade olive oil as better, are 'comparable to cocaine trafficking, with none of the risks'.

In his book *Extra Virginity: The Sublime and Scandalous World of Olive Oil*, Mueller digs into just how this came about. The fault lies first in Italy, the world's most celebrated producer but notorious for decades for ineffective policing of labelling and other frauds in food. Consumers' snobbery and lack of interest in what's behind the label is a problem too: for many countries where olives are not native, Italy *is* olive oil and so we will pay more for that name on the bottle.

Until 2001, European law allowed any olive oil that had been bottled in Italy to be sold as 'Italian olive oil', regardless of where the olives were grown.

The most popular international brand, Bertolli, boasts of its origins in the Italian city of Lucca in the nineteenth century. In 2014, consumers in the United States and Canada mounted a legal challenge to the 'Imported from Italy' and *'Passione Italiana'* phrases on the Bertolli olive oil label, on the grounds that the oil actually came from olives grown and pressed in countries including Greece, Chile, Spain, Australia, Turkey and Tunisia. It had merely been mixed and bottled in Italy.

In 2018 Bertolli's owners, the Spanish company Deoleo, paid US$7 million to settle this case, and though the label still has lots of Italian wording, it merely states the olive oil is sourced in the European Union. Today, Italy produces about 300,000 tons of olive oil a year. Nonetheless, at home and abroad the country sells almost three times as much as this: the regulations still do not work.

Consumers' confusion, or inattention, means that brands are able to use cheap oils in order to produce ones with labels that make them look like premium product. This suits manufacturers, naturally, but it is ruinous to farmers trying to make and sell the proper stuff. And it suits the fraudsters, who, for millennia, have been passing off oil from all sorts of plants as that of olives. Deodorising and cleaning techniques that are used to render seed oil, or even oil chemically extracted from the stones and twigs of olives, produce a very bland oil that mixes easily with pure olive.

Mueller and other lovers of good olive oil accuse the big supermarket brands of a worse sin than mere passing off: corrupting the meaning of extra virginity, a definition of high-quality oil since the Roman emperor Diocletian's rule (and legally enforced since 1960). 'Gentle', 'smooth' and 'not peppery on the throat' are the sort of words used in ads promoting generic extra-virgin oil – and in the United States, particularly, people will buy no other.

The writer Charles Quest-Ritson, a guest on *The Food Programme* in 2007, explained how tastes in premium olive oils have changed: 'The traditional taste – made from fully mature olives – is soft, gentle and almondy. The modern style, championed in Tuscany and Umbria, is for strongly fruity, peppery and bitter oils. There is a place for each of them and many more besides ...' The key to good extra-virgin olive oil, though, all aficionados agree, is that the flavours are vivid.

74

Who knows what olive oil?

I conducted a blind tasting of extra-virgin olive oils a few years ago for a national newspaper. The editor wanted 'the truth on expensive olive oil'. We had a dozen oils, and a panel consisting of an importer, an Italian deli owner and a couple of eminent food writers: the results were so confusing that the article was never published. The tasting was a disaster. The importer went into a rage after he was informed that he'd pronounced his own premium product 'disgusting'; the deli owner chose a bottle of highly dubious 'Italian extra virgin' as his favourite (it had cost £1.99 at the discount store TK Maxx, a tenth of what extra virgin cost in his shop). Both the foodies gave a thumbs-up to the Bertolli brand.

It has become almost impossible for honest retailers to tell when they're being provided with fake oil and, as one sadly tells Mueller, even harder for them to sell good oil for a reasonable price: 'When a customer tries a robust oil, they say, "Oh no, this is a bad oil!" He's become used to the flat taste of the *deodorato*.' As a result, Mueller estimated, when his book was published in 2011, that seventy per cent of cheaper extra-virgin oil sold was a fraud.

HOW TO AVOID BEING AN OLIVE OIL VICTIM

- Never buy any extra-virgin olive oil that does not have a date of harvest or manufacture, and do not buy it if it is more than eighteen months old. The production date is much more significant than 'best before' or 'use by'.
- Why buy Italian? Spain produces high-quality olive oil, and four times as much olive oil as Italy, most of which goes there to be priced-up and resold. While Italy has recently begun to police the industry more actively, with some high-profile arrests, there is no indication yet that the house is in order.
- A cheap olive oil that is green on the plate is very unlikely to be real (it's easy to add the colour). Trust your own senses: you're looking for grassy and peppery flavours, and brightness.
- Look for controls on origin names – the European Commission's PDO (Protected Designation of Origin), DOP (the same in Italian) labels are trustworthy. Genuine organic certification is also a sign that manufacture will have been honest.

LAST DAYS OF THE OIL?

Nothing much appears to have changed since Mueller published his book. He sees the chaos in the olive oil industry largely in terms of honest, hard-working farmers versus slippery businessmen. You may think you could tell the same story for almost any artisanal product we put in our mouths. Industrial production techniques and the supermarket's tendency to strip out quality in order to give 'value' will debase any foodstuff to the point where the producer

Scottish Butteries

7g dried yeast

10g caster sugar

350ml tepid water

125g lard

200g butter

500g strong white bread flour, plus extra for kneading/rolling

1 teaspoon salt

Mix the yeast, sugar and water together in a small bowl and set aside. Chop up the lard and butter and leave at room temperature. In a large bowl mix the flour and salt. Slowly add the yeast and sugar liquid and use a fork to bring it together. Once combined, turn out onto a floured work surface and knead until it is elastic and smooth. Form into a ball, place in a large bowl, score the top with an X, cover in cling film and put in a warm place for about an hour to allow it to prove and expand.

Knead the dough for two minutes on a lightly floured work surface before rolling out to a rectangle shape, about ½ inch (1.25cm) thick. Now you need to layer the fats into the dough. Beat the butter and lard together with a wooden spoon until smooth, then roughly divide into three balls in the bowl. Using your hands, smear one-third of the butter and lard mixture over the lower two-thirds of your rectangle. Fold the top half (without the mixture) over onto the middle third, and then the bottom third up on top of that. Leave in the fridge or a cool spot. After about 30 minutes, repeat the process, but roll the dough the opposite way to how you have folded it. Wait another 30 minutes – chill again – and repeat for the last time, turning the dough again to roll the opposite way.

Roll the dough out to about ¾ inch (2cm) thick and with a knife divide into about 15 pieces, which you can then gently shape into rounds. Put these on a lightly floured baking tray and cover with a large plastic bag or loose cling film. Allow to prove at room temperature for a further 30 minutes.

Heat the oven to 200°C (180°C fan) and when the butteries have finished proving, cook in the oven for 15–20 minutes. Keep a close eye on them from the 15-minute mark. They should turn golden brown.

Remove from the oven and transfer to a cooling rack. Allow to cool slightly before serving with jam or honey and more butter.

Adapted from a recipe at www.scottishscran.com

can feel they have no choice but to commit fraud, abuse animals or take shortcuts to stay afloat.

There are new challenges now for olive trees and olive oil consumers. Droughts in southern Spain, the Mediterranean's largest producer, regularly push up olive oil prices by thirty per cent and more, for a period. Most climate change forecasts predict the region to get even drier as the century goes on. Meanwhile, a new disease is causing devastation across Mediterranean Europe. *Xylella Fastidiosa* is an insect-born bacteria that has come from the Americas and devastates olive trees and a host of other fruiting plants.

While genetic research is underway to counter *Xylella Fastidiosa*, the only known cure so far is holding back the spittle bug that carries it by destroying whole groups of affected trees. This has given rise to furious argument and a refusal among some farmers to even believe in the existence of the disease. In southern Italy, communities have taken to spraying holy water on their treasured trees. But they succumb nonetheless, wilting and going brown – it is death by thirst as the bacteria blocks their irrigation systems.

In Puglia, plant scientist Giovanni Martelli told *The Food Programme*: 'People see these trees, which may have been standing for 1,500 years, as part of their heritage, their history. I see their point; the problem is that if you don't intervene drastically the problem won't be solved, and this will endanger the rest of the trees in the region, the country, perhaps the whole Mediterranean basin.'

The sustainable future of oils and fats in food lies, as much as with any staple, in a return to tradition. There is a place for healthy, sustainably produced vegetable oils. But using animal fats as we once did has many twenty-first-century benefits: we cut food waste, we save valuable crop land here and in the tropics, and we may well enjoy ourselves more, too.

CHAPTER 4

TOUJOURS DU BEURRE

DAIRY

'When our ancestors became farmers, they domesticated cattle and pigs and created a cycle of food and fertility: sun into pasture, pasture into dairy and dairy into meat. A cycle without fossil fuels and waste. It's only in the last century or so that we began to disassemble that system.'

Dan Saladino, *The Food Programme*, 2021

airy has shaped Britain and many other countries in the northern hemisphere. Sheep and cows are the dominant animals of rural Britain, and no farming system has had more impact on our countryside than the husbandry of animals to produce milk, butter and cheese. The classic picture of a traditional mixed farm reflects the needs of cows, with its milking parlour and yard, paddocks and pastures around the farmer and his family's house.

So, too, the economic structure of much of traditional rural Britain. Dairy farming made possible a trade that, for centuries, gave ordinary Britons the cheapest fat and protein available, in town and countryside. Delicious, nutritious milk and 'white meats' – butter and cheese – have long been at the core of our food culture.

The infrastructure shaped by dairy goes beyond the dykes, roads and ditches needed to harbour the wayward half-ton animal that is a milk cow. A major purpose of the railways, when they were first built in the mid-nineteenth century, was to link the dairies with the towns – the milk train is still a term for the first up-to-town service of the day. It puffed its way to town with affordable, nutritious fuel for urban workers. As with potatoes, rice and some other foods we discuss in this book, our modern world would look very different without dairy and its products.

Central to the story of dairy farming has always been the welfare of the animals involved. It can be done kindly or with awful, if unintended, cruelty. The arrival of steam trains first brought a great improvement in the conditions of the cows: before fast transport, many were kept in sheds in the cities, often in grim conditions. By the end of the nineteenth century, milk had become the most important product, by value, of British farming and the industry, even as machine milking began, rural Britain's biggest employer. Even today, there are 12,000 dairy farmers in the UK and the industry employs 75,000 people, perhaps a quarter of all agricultural workers.

So, dairy has for 200 years been crucial, highly political and a very visible part of our nation's food infrastructure. Even now, with sales of vegetable-derived 'milk' soaring, ninety-eight per cent of households still buy dairy products. Over the years, *The Food Programme* has covered the ups and downs of an industry that seems only rarely not to have been in crisis.

TROUBLE IN THE MILK PARLOUR

In the last twenty-five years, dairy has dealt with a series of dire challenges. There has been the mass slaughter of herds because of bovine spongiform encephalopathy (BSE), foot and mouth disease and tuberculosis. Milk producers suffered an outrageous price fixing scam by retailers, which kept milk earnings unfeasibly low, putting thousands of farmers out of business.

Milk producers have suffered from defective subsidy regimes and undercutting from imported products. With good reason, the animal rights lobby has attacked an industrialised system that, to save money, ends the lives of tens of thousands of male calves immediately after they are born. Health scares around saturated fats and lactose intolerance have taken a toll as well. All this while milk remains cheaper than it has been at any time in commercial history – at times considerably less per litre than a bottle of mineral water.

Important to all of this – from failing farms to environmental damage – is the plain fact that British dairy does not market itself well, especially when it comes to the value-added products that make far more profit than can be got from selling fresh milk. When yoghurt arrived on British breakfast tables in the late 1960s, in little milk churn-shaped plastic tubs, it was marketed as 'Swiss-style', even though Ski – the brand – contained as much sugar as ice cream and was made in East Sussex. Consumers will still pay more for yoghurt or crème fraîche whose origins are – or look as though they are – in mainland Europe. For years, half our butter was imported, mainly from Ireland, Denmark and New Zealand. (In 2020, thirty per cent was imported.)

It seems absurd that Britain exports huge amounts of the cheap raw material, milk – 860,000 tons in 2020 – and buys back the expensive processed stuff, in the form of 870,000 tons of yoghurt, butter and cheese. We do export some cheese: about 70,000 tons in the first half of 2021, mainly to the EU. That was down thirty per cent on the previous year. Producers blame Brexit, which has also brought more to add to the paperwork burden. A £150 veterinary certificate is now required every month for exporting dairy products, even those from a small cheesemaker with no cows on site.

THE GREATEST THREAT

Farmers continue to quit the business at an alarming rate: in 2008, at the height of yet another milk price crisis, 1,000 British dairy farmers were said to be quitting the business every year. But these troubles seem small compared to the existential crisis the industry faces today, without doubt the greatest of its 5,000 year history. Dairy and beef farming is now agreed to be the single most damaging agricultural industry in terms of climate change. Raising cattle is a notorious cause of greenhouse gas emissions, much of it methane, which is twenty-eight times more damaging gram for gram than carbon dioxide.

The gases come from the animals' feed and the fertiliser used to grow it, but primarily because of the flatulence and belching entailed by the complex work of breaking down vegetable matter. Then there are the enormous input costs: it can take anything from fifty to one hundred kilos of grass and/or grain-based feed to produce the twelve to twenty-five litres of milk needed to make just one kilo of cheese or butter. The sad truth is that globally cheese is one of the most climate-unfriendly foods that we eat: worse, kilo for kilo, than almost any meat.

A dairy cow eats as much in calories as thirty humans (though weighing only as much as ten). In a year, she will produce 100kg of greenhouse gas emissions. There are 278 million dairy cows at any one moment, worldwide. As a result, dairy farming's climate-damaging emissions are nearly as much as those of global aviation and shipping combined. The business of farming cows, sheep and goats for the wonderful, useful foods that can be had from their milk is as ancient as civilisation. It has weathered so many challenges but can it deal with this, the greatest yet?

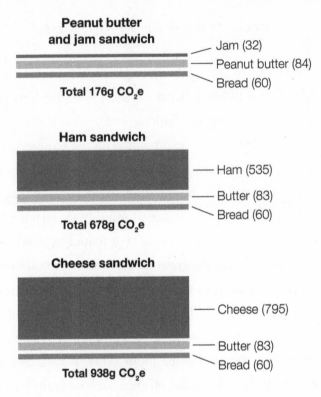

Peanut butter
and jam sandwich

Jam (32)
Peanut butter (84)
Bread (60)

Total 176g CO$_2$e

Ham sandwich

Ham (535)
Butter (83)
Bread (60)

Total 678g CO$_2$e

Cheese sandwich

Cheese (795)
Butter (83)
Bread (60)

Total 938g CO$_2$e

(CO$_2$e = carbon dioxide equivalent, factoring in the impact of CO$_2$ and other greenhouse gases. (Data from S.J. Bridle *Food and Climate Change without the Hot Air* (UIT Cambridge, 2020), based on European farming systems)

ORIGINS

'Mammal' means 'of the breast', and milk from our co-mammals has a history like no other agricultural product. Of the things that demand the coaxing and nurturing of live animals in order to get a regular supply, only eggs have been part of our diet as long. We humans, obviously enough, are familiar with milk from the first day of our lives. The fact that some of us can digest and benefit from the milk of other mammals must have been known by the species of *Homo* long before ours.

Then, as now, animal milk could mean freedoms. First, for women from some of the pressures and timetables of caring for infants and for all people needing quick nourishment, essential nutrition and rehydration. Archaeological remains show that babies were being bottle-fed in Ancient Egypt from as early as 1500 BCE. Only humans steal milk from other mammals and we are fairly choosy about which ones we use. We don't put dogs, cats or monkeys in dairies, though the latter's milk is much closer to our milk than a cow's.

Human milk is 4.5 per cent fat, 1.1 per cent protein, 6.8 per cent sugar (lactose) and 87 per cent water. Seals and whales – hard to get into a dairy – have an amazing 35 per cent or more fat in their milk. Because of the quantity of their milk, and the ease of feeding them, humans long ago chose cows, goats and sheep to keep. They are called ruminants: they chew grass and other vegetation, regurgitating it internally, using their 'rumen'. The average cow produces milk that is 3.4 per cent fat, 3.3 per cent protein and 4.9 per cent lactose.

'Milk always says "Mother" – in our culture, nothing bathed in it can ever be threatening or bad.'
Food historian Margaret Visser

For the proportion of humanity that possesses the enzyme enabling them to continue to digest lactose after babyhood, milk became monumental. It seeped deep into our dreams and fashioned our beliefs. We look up at the Milky Way – a spillage of milk by the Greek goddess Hera – and its galaxies (from the Greek *gala*, for milk). From the Fulani people of West Africa to Hinduism and Norse culture, there are creation myths centred on milk. Romulus and Remus, the founders of Rome, are unusual in that they suckled milk as abandoned orphans from a carnivore, a wolf.

SPILT MILK

Worries over issues from animal welfare to health and now climate change have made milk 'the most argued-over food in human history', as food historian Mark Kurlansky puts it. In the late eighteenth century, there are said to have been 20,000 dairy cows in inner London and this worried people. Some were chained up for their whole lives, eating only grain, brewery waste and rotten vegetables; others were led by a dairymaid from door to door in the early morning, for milking straight into the kitchen pail.

Debates about hygiene, disease and the cows' wellbeing began: they show no sign of being resolved in the twenty-first century. Modern dairy's enemies begin with the accusation that we have, as with chickens, turned a good, mutually beneficial relationship with a sentient animal that is capable of complex emotions into a deal no more humane than exists between man and machine.

Breeding and 'zero-grazing' feeding systems – where high-energy, grain-based fodder is delivered straight to the cow's stall – have turned the modern cow into an amazing milk-producer. Ten thousand litres a year is normal from a standard Holstein-Friesian: forty years ago, the yield was half that. A beef cow feeding its calf will provide about 1,000 litres.

This extraordinary efficiency is what keeps milk so cheap in the supermarkets (though without giving much profit to the farmer). But the cost to the millions of animals is terrible: lives that last three years instead of eight or more in an organic dairy and, for some, a grassless existence in a shed not so unlike those that appalled animal-loving Londoners 200 years ago.

At the heart of the problem is a fracture in the ancient way of using cows in agriculture: a system that was efficient, self-contained and as humane as possible. Simon Fairlie, a farmer and author, explained to *The Food Programme* in 2021 how the system works on a traditional English mixed farm. 'The cows are at the heart of it. They produce about 4,000 litres a year from land that you couldn't cultivate, it's rough, hilly grassland. That produces milk, yoghurt, cheese – cheese is the way you preserve what you can't eat immediately, and when you make cheese, you end up with whey, which has still got a substantial amount of goodness in it. That goes to pigs – they love it. And then the cows produce seven to ten tonnes of manure a year, which we compost. The pigs as well are producing manure and we plant potatoes on the land where the pigs have been.'

A side-product is beef for eating. It comes only from the animals that are not productive for the dairy. Fairlie considers this not just efficient but a morally justifiable use of livestock. 'The beef industry should really go back to using dairy cows or what are called dual-purpose cows.' The herd would be fed mostly on the pasture available. 'It's really inefficient feeding grain to any animal,' says Fairlie. Nevertheless, in some parts of the world, beef farming has been entirely separated from dairy, and the animals fattened on soy and other grain-based animal feeds. This in turn drives deforestation in Brazil, Argentina and other soy-producing countries.

EXTENDING THE LIFE OF MILK

The arguments about dairy to which Mark Kurlansky refers rise chiefly from the various ways we've turned milk into money. It is a great staple, providing protein, fat, sugar and some useful minerals, like calcium, at a very low cost. But it doesn't last; in fact, it will go off very quickly compared to most animal products. Dangerous bacteria start to multiply in fresh milk within minutes, if given warmth and an unclean receptacle.

For centuries, methods of making it safe – or look that way – have proliferated. In the modern era, milk has been homogenised, pasteurised, condensed, powdered and evaporated, then sold in tins, bags and bottles for many things from baby formula – still sold in some countries as cleaner and more civilised than human milk – to chocolate bars. Stabilising milk's fats as butter and then salting it to preserve it brought about one of our first luxuries, a key element in northern Europe's richest cooking: '*Donnez-moi du beurre, encore du beurre, toujours du beurre!*' cried the early twentieth-century chef Fernand Point, a father of modern French cuisine.

Very early on, humans started boiling milk, which gets rid of some bacteria. (The ancient Assyrians treasured the nutritious skin that would appear on the top of the heated milk.) But a more effective way of extending the life of milk and adding value was to exploit one of its many shape-shifting properties. Butter is made very easily from milk – we used to do it as children simply by carrying a jar of milk round in our pockets. Warmth and motion do the job. Salting it will then ward off bacteria. The Sumerians were shaking cream in a goat skin to make butter around 4,500 years ago: they had a goddess, Ninigara, in charge of it.

Milk goes off because the action of bacteria from the environment around it turns it sour and smelly. But many of these bacteria have useful effects. Yoghurt is a central food to many Middle Eastern cultures and it too needs motion to make it, after the adding of the bacteria, usually the remains of the previous day's batch. These bacteria produce the acid that makes the yoghurt tart and give it its texture.

In ancient Persia, a transportable sun-dried, fermented yoghurt emerged. It is still around, as *kashk*. The chef Yotam Ottolenghi treasures it as a tart seasoning or a thickener for soups and stews, while in Lebanon and nearby countries the yoghurt is mixed with ground wheat and dried to make a kind of biscuit. Yoghurt – the word is Turkish – was of course a dubious, foreign thing for the British until the 1970s. 'I'm not eating milk that's off,' cried many consumers.

Add a little acid and shaken milk will begin to coagulate and then separate into soft lumps – curds – and a watery substance – whey. You can eat and drink them – Little Miss Muffett did just that. Curds, washed and salted, become cottage cheese, while whey is often used to feed pigs. But with more time and some interesting bacteria, they are the beginning of many good things, not least cheese.

The usual explanation for the beginning of cheese is that herdsmen would carry milk for convenience in a waterproof sack made from an animal's bladder or stomach, only to find that, after hours of travelling, the milk had turned solid, and tasty. What had happened, and all praise to the first human to work this out, was that a mixture of acidic enzymes we call rennet, present in ruminants' stomachs, had curdled their milk, enabling its molecules to fuse. This produced curds.

The first recorded cheesemaker lived way back in the mists of Greek legend, perhaps 3,000 years ago. He was the one-eyed giant demi-god Polyphemus, who milked ewes and made curd cheese (what we would call cottage cheese), until he was tricked and blinded by the Greek voyagers in Homer's *Odyssey*. But cheesemaking is far older than that: archaeologists working on a near-6,000-year-old salt works in North Yorkshire in 2021 found evidence that the workers there were processing milk into solids and storing it in pottery.

A TWENTY-FIRST CENTURY CHEESEMAKER

Cheese and its making are ripe with ancient skills, some of them the stuff of legend. Many of the most prized British cheeses come from hallowed traditions centuries old: the blue-veined Double Gloucester, a ploughman's lunch staple, is said to have first been made in 1498. Artisanal cheesemakers are rightly protective of their cheese's names and time-honoured techniques. But, as the story of Feltham's Farm's cheeses shows, you can design and market a cheese from scratch, using off-the-shelf ingredients, even among the pressures and uncertainties of a twenty-first-century dairy.

Marcus Fergusson and Penny Nagle moved from London to the Somerset countryside in 2015 with their three small children.

They wanted to farm but their new home, Feltham's, had just twenty-two acres – too small for conventional agriculture, which tends to go big nowadays to be viable. So they decided to make cheese. But what sort? Somerset is full of good cheddar makers and others doing local variations on French classics like Brie and Camembert.

'We said we've got to produce a cheese that is not commonly available or not available at all. We looked at types of European cheese you don't get over here that often. We thought, okay, stinky, washed rind, French cheeses like Époisses, they're quite fun. But how could we put a British slant on that?'

Designing a cheese today is a matter of assembling the building blocks, says Marcus: the milk, a rennet, salt and the culture – the bacteria, yeasts and moulds that define the cheese's character. These can be bought from one of two big Danish companies, between them responsible for the cultures behind eighty per cent of cheeses in the world today.

Marcus and Penny started with a basic Camembert recipe, taking out the culture that makes the bloomy white rind. In went some *Geotrichum candidum* to give a sort of 'wrinkly brain texture'. Then he added some blue cheese culture, *Penicillium roquefortii*. Finally, because Époisses is washed in brandy to encourage the growth of 'certain funky bacteria', the Fergussons decided to finish the cheese with a dip in local Somerset ale.

'In a way, the milk tells you what it wants to be,' says Marcus Fergusson. 'It and the environment you're in have a lot of input. You can play with the majority of the elements but, truth be told, after mixing all the different cultures, you invariably come up with something you didn't expect.'

93

The cheese Feltham's Farm came up with after some experimenting is a real character. It is the kind you're nervous with; slightly sweet and very smelly when ripe, it needs a sealed box in the fridge. An early review called it 'a vicious little cheese', a line Marcus likes. With an attention-grabbing name – Renegade Monk – and a steampunk label, they started trying to sell it in local shops.

The success was beyond all hopes. Within the year, cheese buyers were queuing up, not least because just five months after production started, Feltham's Renegade Monk won the gold medal for Best Soft Cheese at the 2017 Global Cheese Awards. A string of further gongs and plaudits followed, including stars from the Great Taste Awards and the British Cheese Awards in 2020. Renegade Monk is now exported to Canada and Hong Kong, and Marcus and Penny, having enlarged their dairy, are producing three more soft cheeses, including a milder blue cheese and England's first Spanish-style *queso fresco*. Their production is still tiny, though, hit by coronavirus lockdowns and all the other recent stresses on rural businesses.

How have climate concerns impacted production? It takes around seven litres of milk to produce one kilo of Renegade Monks (relatively little: the harder the cheese, the more milk is required). Unlike some in the dairy industry, Marcus is not in denial about the climate cost of his product, and from the beginning of the business he and Penny have worked 'to tread as lightly as we could'. The farm is certified organic and the electricity, heating and refrigeration are supplied by solar panels and a ground-sourced heat system. Their small herd of pigs eat the whey from the cheese making, which otherwise would be discarded. The farm van is electric and over 1,500 trees have been planted in four years. 'I have not run the numbers since our dairy was upgraded but I'd say we're not carbon neutral, but carbon negative.'

GREENING THE MILK

The carbon cost of the milk the Fergussons use is a rather more complex issue. Feltham's Farm is supplied with milk by two local organic dairy farms. One is Godminster, which also supplies the big cheddar firm, Wyke. It is, officially, the 'best large organic farm in Britain', as designated by the Soil Association in 2021. Its owner, Richard Hollingbery, bought the 100-year-old dairy farm outside Bruton in Somerset in 1993. Since the beginning, he says, his watchword has been 'nature repays those who treat her kindly'.

Godminster gained its organic certification in 2000. It means that the cattle – beef and dairy – eat a minimum of seventy per cent grass, that no artificial chemical inputs are used on the land and no antibiotics are used in their treatment. The pastures are not fertilised and when seed is planted it is done on the 'minimum-till' system – as Richard puts it, the turf is just scratched up. Traditional heavy ploughing turns the topsoil over and releases trapped carbon dioxide. Minimum tilling, which some cereal farmers are now using, conserves the soil structure by only going 15cm deep. It also uses much less in diesel.

It is about as 'good' a farm as is financially possible nowadays: 25,000 trees have been planted in the last twenty-five years and Richard Hollingbery continues looking for more ways to reduce energy use and improve in other ways: he is currently trying to see if his milk customers will accept compostable packaging, rather than single-use plastic.

The farm is also part of a scientific project turning the methane produced by the slurry pit – where the cows excreta goes when they are in the yard – into bio-fuel. And, by using sperm selection science, the farm has produced no male calves from the dairy herd for eight years. Even before that, and unlike in many industrial farms, Godminster's bull calves were not killed at birth but raised for veal. But this was uneconomic and, of course, another producer of climate-altering gases.

One of the great but rarely admitted horrors of modern industrialised dairy is the slaughter of huge numbers of male calves shortly after they are born. They are too expensive to raise without their mothers' milk and, since dairy breeds often do not produce good meat, their carcasses simply end up in landfill. But in Britain, at least, the problem is now being addressed.

Choosing only X-chromosome-bearing sperm for artificial insemination is one technique being perfected: it has allowed some farms, like Godminster, to ensure only female calves are born. Driven by customer complaints, the big milk industry has acted too. Arla Foods, responsible for twenty-eight per cent of British dairy farms, introduced rules in January 2021 banning its 2,300 farmers from killing calves under eight weeks – a move first suggested by the farmers themselves. Developments in the genetics of dairy cattle will, in the future, allow more of their male calves to become beef.

Despite the attention being paid at Godminster to the farm's sustainability, when we spoke in late 2021, Richard Hollingbery was feeling flummoxed by the task of dealing with his dairy herd's greenhouse gas emissions. 'We're trying to write a new sustainability plan but the biggest wall I've come up against is that there's no agreed formula for carbon auditing. Nothing from government and at least three different systems, depending on who you ask.'

He wants to look at the farm's total greenhouse gas expenditure, from the cost of the server farms where its computer data is stored through to the landfill problem of his customers' waste disposal habits. 'But it cannot be done.'

On the specific issue of the cows and their burps – they burp more than they fart, he says – Hollingbery feels his end of the dairy industry has been very unfairly attacked. 'When you see horror-inducing headlines in the *Guardian*' – a recent one said that thirteen dairy companies produced more greenhouse gas than the rest of the UK's entire emissions from everything – 'you have to read down to see that they are talking about multi-national corporations with mega farms, 80,000 cows in a two-square-mile area in New York state all being fed on soya pellets.'

Five per cent of the calories each cow eats are belched back out as methane – a greenhouse gas that is twenty-eight times as damaging as carbon dioxide.

FIGHTING OVER THE DETAIL

'The Facts about British Red Meat and Milk' is a combative 'myth-buster' paper put out by the National Farmers Union (NFU). It states that the United Nations Intergovernmental Panel on Climate Change has acknowledged that British dairy is far more efficient, in terms of emissions, than the global average – producing forty per cent of what other countries and systems do per cow. That is in part because currently Britain imports less soya for feed than do other European dairy countries.

Grass-based dairy also sequesters carbon, taking it out of the atmosphere, since the plants themselves consume it. The NFU paper makes the important point that while methane is much more damaging, pound for pound, than carbon dioxide, it degrades and disappears quicker in the atmosphere. CO_2 remains for centuries.

It is arguable, though, whether traditional grass-based dairy herds are always the better option; recent science suggests that modern 'semi-intensive' dairy farming that uses feed efficiently and disposes of faeces effectively may be less damaging to the environment. Although there are many forward-looking farmers, like Richard Hollingbery, the bulk of the British dairy industry seems determined to fight this battle not with changes to its core practice but by arguing down the science and making the point that the positive things about British farming – jobs, environmental protection of the land, history, aesthetics – outweigh the negative. As with so many players in the great climate crisis argument, the approach often seems more defensive than constructive.

Kite Consulting is one of Britain's leading analysts of industrial agriculture and its markets. Some of its staff see the dairy industry as unfairly under attack over the climate crisis, from people with 'agendas against ruminants', including the BBC, 'the vegans' and the fossil fuel industry – which in turn is seeking to deflect from its own notoriously bad record on greenhouse gas mitigation. The global science and statistics are unreliable and abused, says Kite Consulting's managing partner John Allen, and the ordinary public are left utterly confused.

Speaking for Kite Consulting, PR expert Chris Walkland says it is all about taking a simple message to the public. 'Tell them to simply go to Cumbria, go to Devon, go to Wales, go to Scotland, and look out the bloody window. See the countryside, that is kept by animals, and we can tell them that 300 cows are normally surrounded by 300 acres of fields and hedges and trees and everything, absorbing that carbon. Then go to the middle of a city and out the bloody window there's nothing to absorb that carbon.'*

But Rachael Madeley Davies, head of sustainability at Kite Consulting, knows that the dairy industry needs to do more than this. She is working on the development of different greenhouse gas auditing systems to help address the problem Richard Hollingbery at Godminster has in trying to measure all the factors in his business. Before long, she believes, a milk bottle will carry data on the carbon footprint of the farm it comes from, which will in turn bring financial benefits to farmers, especially if the government introduces a carbon tax. Already, Tesco is working on a scheme to measure farms' greenhouse gas emissions and requires them to comply with a certain maximum level of carbon per pint in order to get onto the supermarket's shelves.

With scientific developments, dairy and beef farms can make themselves carbon negative, increasing biodiversity and taking more greenhouse gas out of the atmosphere than they put in. Rachael Madeley Davies says, 'We are at red alert on the climate. It's good to change, and we have to, because there's a lot of issues heading dairy farmers' way.' She mentions water use and the problem of

---

* 'How dairy can be part of the climate solution', Kite Consulting podcast, 20 August 2021

the release of polluting ammonia through urine. 'We can adapt, we're working hard at it, but it's going to be very difficult for a lot of people in the industry.' Nonetheless, one big cattle farm in Australia, Wilmot, has sequestered enough carbon through new grazing and land management techniques to enable it to sell carbon credits. In 2021, Wilmot did a deal worth half a million dollars in carbon credits – which allow a business to pay another to offset greenhouse gas production on its behalf – with the computer and software giant Microsoft.

Ornua, processors of Kerrygold butter, Pilgrim's Choice cheese and own-brand dairy products for UK supermarkets, said in 2021 that it hopes to achieve a twenty per cent reduction in its emissions by 2030. Other changes have also helped: reforming its supply chain has reduced the business's energy usage by thirty-six per cent over the past three years. Ornua and Tesco have reduced plastics use, too, by changing the design of butter and cheese wrappers. While many in the dairy industry fulminate at being made scapegoats by a lobby that, they say, wants to end all animal farming, behind the scenes real change is happening.

The future, many analysts believe, lies in bio-technological change, not least in cows themselves. Putting a small amount of seaweed in feed can reduce the animal's methane production by sixty per cent or more: trials are continuing. Genetic work to improve cow's health will keep them productive: an ill cow produces just as much gas as a healthy one but much less milk. Cutting the size of the worldwide herd while maintaining milk production levels is thus a possibility, especially if the technology can be given to big milk-producing countries like India, where cows at the moment are much less efficient.

'Cow's milk without cows' will be in the shops by 2023, newspaper headlines were promising in late 2021. An Israeli start-up was reporting that it had raised investment for a project that would insert DNA into fungi or other plant microorganisms, enabling them to produce casein and whey, the proteins in milk. Then the company, Imagindairy, will add water, plant-based fats and sugar to make a pint of milk using one per cent of the land and ten per cent of the water used by dairy cows.

You may imagine that investors' enthusiasm for this and other schemes is driven less by worries about the toll of dairy farming on the planet and more by the massive growth over recent years of the non-dairy milk market. Plant milk, made from almonds, oats, soy and other sources is forecast to be turning over more than half a billion pounds annually by 2025 in the UK alone.

AT THE CHURN - WOMEN AND BUTTER

Women have always led the hard work of milk processing – you could argue that it was forced upon them. The word 'dairy' comes from the Middle English word *daie* for 'female servant'. Dairymaids, whether leading a cow door-to-door through the streets of London, working in the milking parlour or churning butter, occupy a place in the nation's imagination and folklore. In Thomas Hardy's novel *Tess of the d'Urbervilles*, dairymaid Tess Durbeyfield is a symbol of beauty and rural innocence – soon to be corrupted.

One of the best murder stories from the Bible's Old Testament involves a strong woman and a gift of butter. Jael was a Kenite, an ancient tribe of nomads in what is now northern Israel. They

'But the milk itself should not pass unanalyzed, the produce of faded cabbage leaves and sour draff, lowered with hot water, frothed with bruised snails, carried through the streets in open pails, exposed to foul rinsings, discharged from doors and windows, spittle, snot and tobacco quids and trash chucked into it by roguish boys and, finally, vermin that drops from the rags of the nasty drab, that vends this precious mixture under the respectable denomination of milkmaid.'

Tobias Smollett on the morning milk delivery in London, *Humphrey Clinker* (1771)

102

would, as herders, milk their camels and goats and trade the produce. Jael took her opportunity when the Canaanite general, Sisera, an enemy, appeared at her encampment demanding food and a place to rest (and, some translations suggest, sex with Jael).

Jael gave the soldier a bed in her tent, brought him 'butter in a lordly dish' and then, when Sisera was asleep, took a tent peg and drove it through his temple with a hammer, killing him. As my mother said when I brought this story back from Sunday school, 'I don't think it would have worked with Stork margarine.'

Lore in many dairying cultures had it that women were more suited to the job: the cows preferred them. Over the generations, this consigned many women to long hours in the milking parlour. A herd of twenty cows needing milking twice a day was six or more hours of work, by hand. It is not much quicker today using machines, though less demanding on the arm muscles. Then the milk had to be processed.

Clearly the labour offered opportunities for many women, but family members would have worked unpaid in this strenuous and occasionally dangerous labour (half a ton of bad-tempered cow can hurt), close under the farmer's watchful eye. Milk was hugely important to the family and to the farm's economics: a cow that lost her productivity was a disaster and often a woman would be blamed.

Accusations of witchcraft appear again and again in tales of dairy making; blood in the milk or curdling, the souring of butter or its failure to emerge were all blamed on malign women, or sometimes merely on menstruation. Superstition was part of dairy parlour life – with horseshoes and rowan tree twigs hung on the cows' stalls to ward off bad luck. Hedgehogs, hares and nightjars – perhaps introduced by witches – have all been accused

103

of bringing milk-spoiling bad luck or even of stealing the milk from the cow's udders.

A normal day for a dairymaid started at 5 a.m., loading the morning's milking into the butter churn. She would sing a churning song, to get the rhythm as she swung the handle:

> Come, butter, come.
> Come, butter, come.
> Peter's standing at the gate,
> Waiting for a buttered cake.
> Come, butter, come.

If the butter did not come – which means separate itself from the milk – the blame fell on the dairymaid. A simple mistake like failing to say 'bless this work' on entering the dairy could be the problem, it was thought. Women with hot or clammy hands were no good at the work. It was not until the modern era that science worked out that acidity in the milk or problems with the cows are the actual – and more reasonable cause – of butter that would not come.

The relationship with a cow you've milked well and gently, as anyone who has done it for a while knows, can be a warm and happy thing. The best butter of all was made from the last, creamiest milk of a milking, carefully coaxed from a compliant cow with kind fingers – a milk called 'the strokings'.

Processing the butter for storage and sale occupied yet more time in the dairy. After churning, when the butter is kneaded to remove whey and make it homogenous, brine would be added. Well-wrapped salted butter would last for months at temperatures as high as seven or eight degrees. Today's is only about two per cent salt – a record from the early fourteenth century suggests that one pound of salt was added to every ten of butter. In search of the cool, some Northern European butter-makers buried their product in wooden tubs in peat bogs. This produced a great flavour and they buried some for so long that they would plant a tree to mark the spot. Archaeologists in Ireland who have dug these up say that, centuries on, the contents are whiteish and well preserved, though more cheesy now than buttery.

PURIFYING THE LIFE OUT

Many of the modern improvements in milk came about to save lives. Though the relationships between dirt, warmth and infectious disease were hardly understood, milk had been regarded as suspicious by doctors since the seventeenth century. It was obvious how wrong it could go, very quickly. In the cities cows were often fed chiefly on swill – the waste mashed grain from breweries – sometimes so much that the milk smelt of alcohol. So things were added to make it look better and fresher.

The snail juice that Tobias Smollet mentions on page 102 was used to make milk look more frothy and it was hardly the worst ingredient that has been added to milk over the years. Chalk, flour, plaster of Paris, sheep's brains and even highly poisonous lead-based white colouring were also found when the Victorian scientists started policing the urban milk business. On top of that, when tuberculosis was first understood and investigated in the 1880s, it was found that most of London's cows had it.

In the 1860s, Louis Pasteur discovered that heating organic liquids – like wine and milk – to around boiling point would destroy microorganisms and deactivate enzymes. Nowadays, most milk is taken to 72°C for about fifteen seconds before it is cartoned or bottled. Pasteurisation saves lives. It is one of the processes that shaped the modern world, but it is not the greatest way to treat milk if you want nourishment from it. 'It is an irony that, having identified the "bad" bacteria that soured milk, [Louis Pasteur] invented a way of preserving the liquid by killing *all* its microbial life,' writes historian Carolyn Steel. She compares pasteurisation to other blanket solutions, like pesticides, and blames it for some of the problems we now face as a society that has become too clean-living, wiping out natural bacterial systems that actually benefit us.

The bacteria in raw milk have served humans well, as have the yeasts that give us beer, wine and yoghurt. They not only provide taste but also assist with preservation. The horror show that was milk in the early industrial age served to make sure that what we now drink is identical, dull and lacking in some of its best properties – not least vitamin C, which pasteurisation also destroys.

Raw milk's 'probiotics' have been linked (that word again) not just to better functioning digestions but to the treatment of asthma, eczema and allergies. It is thought the natural bacteria in milk help build our immune system. Thus, untreated fresh milk has come back to consumers as a health product, especially for the audience that is concerned with repopulating its gut biome with fermented drinks and foods like kefir and kimchi.

Mainstream science remains unconvinced. In 2017, the Centers for Disease Control and Prevention (CDC) in the United States issued a stern report linking outbreaks of campylobacter, salmonella and listeria poisoning to a rise in raw milk consumption.

They said it was causing ninety-six per cent of the outbreaks linked to dairy products in the States – a lot, when you consider that at the time only 3.2 per cent of Americans drank raw milk.

It is in traditional cheesemaking that raw milk matters most. Some cheese eaters turn up their noses at cheese *not* made with raw milk – because traditional cheese, of course, uses those specific bacteria that come with the environment and place where the cows were milked. Carolyn Steel cites Cantal, made in France's Auvergne. Made from the milk of Salers cows, it has been enjoyed for a long time: the historian Pliny wrote, in the first century CE, that it was the 'most appreciated in Rome'. The key, it turns out, is the old wooden vats in which the fresh Salers milk is kept. Never washed, they are alive with the right microbes to make waxy, pungent Cantal.

Such lore does not work with people who read US government health reports. After a child died from a listeria outbreak in a Scottish pub in 2016, Scotland's environmental health police were encouraged by the government's food standards authority to close down a famous traditional cheesemaker nearby. They then encouraged all Scotland's other cheesemakers using raw milk to change to pasteurised or shut up shop. This, it was pointed out, took place while supermarkets in Scotland continued to stock raw milk cheese imported from Europe.

'I SCREAM FOR THE ICE CREAM'

From its role as a staple, milk products began to become treats and luxuries in the industrial age. Itinerant Italian ice cream vendors were a regular feature of mid-Victorian London's streets. Their cry *'gelati, ecco un poco!'* led to them being called 'hokey-pokey men' and ice cream was known as 'hokey-pokey' well into the twentieth century. Contemporary with the arrival of this treat in the crowded cities were great plagues of diphtheria, scarlet fever and typhoid.

It is now thought that ice cream may have been a key vector for their transmission. At that time, people believed that cold was lethal for bacteria, though in fact many species thrive well below freezing. By the 1920s, it was understood that an ice cream mix – which included egg and added sugar – might be a more effective medium for growing bacteria than milk or cream themselves. Since heat kills most bacteria, pasteurisation became legally required, first for ice cream and then for most fresh milk. A simple test for phosphatase, a chemical destroyed in milk when its temperature is raised to 72°C, could show very quickly if pasteurisation had been successful.

Thus rehabilitated, and with the advertising age beginning, factory-made ice cream became the most adored dessert – in milk-eating societies – of the modern household. As the middle classes of wealthy nations acquired electric refrigeration, what better way to show off the shiny new gadget than with a brick or a half gallon tub of ice cream?

The food historian Margaret Visser sees twentieth-century America's relationship with ice cream as obsessional – a national addiction and a symbol of national plenitude. In the 1920s, new immigrants arriving at Ellis Island off Manhattan were served a plate of ice cream as they formally entered the United States. When, in 1942, the battlecruiser USS *Lexington* was sunk in the Battle of the Coral Sea, the crew raided the freezers so they could sit on deck gorging on ice cream as they waited until it was time to jump overboard. Meanwhile, both the Japanese and the British banned ice cream in the Second World War as an inappropriate luxury.

So important was ice cream considered to the mental health of the young men sent off to the Korean War in the 1950s that the US Army put logistics in place to get frozen flavoured cream, fruit flavouring and egg to all troops, even deep in the jungle, three times a week.

Dulce de leche

Not many foods are new or better because of modern, mechanical means of preserving them. Safer, yes, but more delicious? Well, there is *dulce de leche* – sweetness of milk – the gorgeous, copper-coloured soft caramel spread that Latin American cooks make from tins of condensed milk. This was one of the first tinned milk inventions, made by adding sugar and boiling away most of the water. First produced in the 1850s as a safer food for children, condensed milk was provided in huge quantities as emergency rations to Union soldiers in the American Civil War. After that it became a staple of US dessert cuisine.

Classic *dulce de leche* is made by boiling the unopened tin in a pan of water for 2–4 hours. But as Catherine Phipps writes in *The Pressure Cooker Cookbook*, you can do it in 20 minutes in a little water in a pressure cooker – 30 minutes if you want it darker and more caramelised. ('Not risky at all,' she promises.) Add vanilla, a little salt and spread the paste on toast, serve on top of ice cream or use as a cake filling.

Today, as all ice cream eaters know, most of the product has travelled far from frozen cow or ewe's cream. It is a technological jigsaw, an assembly of gums, starches, fat powders, colourings and flavourings collected from seashore, cornfield and forest. Most people don't mind the trickery: food that masquerades has always been a delight. As Visser says, in Anglo-Saxon countries ice cream has always been the 'apotheosis of milk': a luxurious treat that marks the finest moment in the marriage between human ingenuity and the basic nourishment with which we all started life.

LANDS OF MILK AND HONEY

You cannot help but think we've managed the business of milk and its many products poorly in the modern age. Government in Britain has attempted to control milk production and prices off and on since the late nineteenth century, yet even the price of a pint is not yet fixed in a way that ensures a dairy farm's viability. It is not easy. Balancing protection for dairy farmers while also ensuring the right amount of milk left the farm gates was never a success. But doing away with the Milk Marketing Board and leaving milk supply to the market, as Britain did in the 1980s, led to disaster for many farms as the retail price collapsed.

'Milk lakes' and 'butter mountains' – massive stockpiles of unsold produce caused by farm subsidies in the European Union – were an issue in the late twentieth century. One way of dealing with the excess was to export the European milk – after heat treatment to make it 'long-life' – at prices well below cost around the world, an enterprise that managed to destroy the dairy industries of Jamaica and several other much less wealthy countries.

The truth is that British dairy processing – despite the successes of a great tribe of new artisanal cheesemakers – seems not to be very nimble. How else do you account for the market dominance in Britain of Lurpak, a butter imported from Denmark? No British butter maker has ever managed to rival it. Yet. It is simply a matter of the right culture, as the Feltham's Farm story shows.

What is a modern British dairy farmer to do? They are certainly right in their complaint that for decades supermarkets have kept milk prices unfairly and needlessly low, often below the cost of production. In 2007, three major ones were fined millions of pounds for operating an illegal milk price-fixing deal.

Despite the fact some chains have now entered into new contracts to guarantee base prices, it's hard not to conclude that an awful lot of the problems discussed in this chapter – from disposal of unwanted calves to how to address the dairy industry's role in climate change – start with the cost of a pint (55p in Tesco at the end of 2021). As Richard Hollingbery says flatly, 'Good dairy farming is about a good milk price.' Happy cows need the same.

CHAPTER 5

SWEET MADNESS

SUGAR

'Sugar is not just a problem. It's part of a culture,
part of a cuisine to celebrate all the great seasons
of the year and events in our lives. So what
should our relationship with sweetness be?'

The Food Programme, 2013

The desire for sugar is hard-wired into humans. A 240ml cup of our milk has 17g of sugar in it – a saturation rate about halfway between that of cow's milk and of squeezed orange juice or Coca-Cola, though breastmilk is more complex. When a baby starts to suckle successfully, endorphins are released in its brain: a feedback loop, the association of pleasure and comfort with sugar, is established. The breastmilk is also full of vitamins, minerals, antibodies, proteins and compounds that get gut bacteria started. But it is the sweetness that drives the urge.

Child development expert Jacob Steiner was the first to spot that babies, then thought to be expressionless until around a couple of months old, licked their lips and gave a 'slight smile' at the taste of sugar at one week old. From these beginnings, a form of addiction begins; anyone who's found a toddler with their fist in the sugar bowl knows the strength of the compulsion. But it is worth remembering that milk's fat and its sugar, called lactose, is crucial to the developing baby. The still semi-formed brain and body needs lots of easily obtained energy to grow.

With ample reason: no food in this book is as reviled as sugar is today. Health writers and food and cultural historians have taken this on, even if consumers remain blissfully – or wilfully – in denial of sugar's many problems and hazards. From the health and the history sections to the cookery shelves in a high-street bookshop, you can see the pariah status of what remains the world's most lusted-for ingredient. *How Sugar Corrupted the World*, a 2017 book by the historian James Walvin, carries the subtitle 'From slavery to obesity'. Over on the medical shelf a classic, Professor John Yudkin's 1972 *Pure, White and Deadly*, sits beside Dr Robert Lustig's newer bestseller *Fat Chance: The bitter truth about sugar*.

There are not many books that will make you feel good about sugar. On the cookery stand, for every recipe book about cooking with it – new in 2021 was pastry chef Ravneet Gill's *Sugar, I Love You*, a welcome exception to prove the point – there's five more offering no-sugar baking, including *I Quit Sugar* and *The No Sugar Diet*. Both being literally unachievable, given that there's sugar present in just about every natural food known to mankind.

All the same, no food is quite so demanded and adored as sugar. There's a whole alphabet of treats beloved by humans. Take the Ms alone: marshmallow, meringue, macaroon, maple syrup, melon, marmalade, marzipan, muffins and mango. And then there are Mars Bar, Milky Way and Maltesers.

These are all 'high in sugar' – in the UK that means 22.5g, four teaspoons or more per 100g. The melon and the mango's sugar content is intrinsic and natural; their effects inside our bodies apparently different from the added sugar of the processed products, most of it way beyond what is necessary for taste and structure. None of the processed products would function in the way we know if the sugar was removed. What on earth are we health-conscious sugar-eaters, conflicted and confused, to do? And how on earth did we get to this point?

SUGAR ALCHEMY

Humans have done some wonderful and terrible things because of their addiction to sugar. Looking at the first, shorter list takes you deep into our culture and our history. Sugar's prime attribute is as a cheap source of instant energy but it has other unique abilities. Why, for a start, does your bread go that pleasing brown when toasted? You are experiencing the Maillard reactions, named for the French physician who in 1910 identified the chemical changes that occur when proteins and sugar encounter heat while he was researching the function of the human kidney.

The heat from the toaster breaks down the sugars on the surface of the bread. They then combine with amino acids in the bread to produce a host of different flavour compounds. The same is going on when we roast coffee beans or meats. The reaction principally uses glucose, one of the two molecules that make up sucrose – standard sugar – the other is fructose. Maillard's reactions happen at around 154°C, and also produce colour. Bagels and pretzels get their glossy brown surface because they are dipped in an alkaline wash before cooking, accelerating the reaction and increasing the production of the brown pigment that tells us they're cooked just right.

Caramelisation is the most delicious of sugar's heat transformations. It is not just about colour and sweetness but also the breakdown of compounds to produce vivid and rich flavour. Try pan-grilling a spear of asparagus, using heat to break up its natural sugar. Most sugars begin to caramelise at about 160°C, though the individual elements, sucrose, fructose and glucose, all have different caramelisation points. Sucrose is made of linked fructose and glucose molecules – complicated chains, in essence, of atoms of carbon, hydrogen and oxygen.

At around 180°C, caramelisation's best flavours start to appear – which is why when you roast a joint of meat, the outside temperature has to rise much higher than it does at the core, if you want it to brown. Inside the meat a mere 60°C is enough to break down fat and tendon, and kill bacteria.

The higher you turn the dial, the darker the surface of the food becomes. The colour is also a danger warning. When the toast is dark brown or black and bitter tasting, you've used that thermal energy to create acrylamides through a denaturing effect known as pyrolysis – coming from the words 'fire' and 'separation' in Greek. Acrylamides are carcinogenic: the black and the bitter are telling you that burnt food – or burnt coffee – is not good for you.

FIRST CAME HONEY

On a cave wall at Arana, near Valencia in Spain, there is an extraordinarily vivid depiction of a human up a tree collecting honey, outsized bees buzzing around the thief. It is 8,000 years old, far older than any other depiction of humans getting sustenance from animals without trying to kill them. Our Palaeolithic ancestors would have found the honeycomb by tracking wild bees back to their hives. The Hadza of Tanzania, one of the last remaining people who can be described as hunter-gatherers, find honey by whistling to the honeyguide bird, a type of woodpecker. Its Latin name explains its role: *Indicator indicator*. The bird leads them to a hive, knowing that the humans will leave honey traces and bee larvae for it when they have taken what they need. Dan Saladino saw this happen for a *Food Programme* episode in 2017, recording the yells of the Hadza man who climbed thirty feet up a baobab tree to smoke the bees out of their home and got copiously stung as he collected the honeycomb.

The writer Washington Irving witnessed a bee hunt on the American frontier in 1832, in what would become Oklahoma. He wrote that the Native American honey-gatherers he saw rarely encountered honeybees until their land was invaded by the Europeans. They considered the insects to be a warning, 'the harbinger of the white man' – wild colonies that spread from domesticated ones brought by white civilisation. There was a bonus, though. 'The Indians with surprise found the mouldering trees of their forest suddenly teeming with ambrosial sweets, and nothing, I am told, can exceed the greedy relish with which they banquet for the first time upon this unbought luxury of the wilderness,' wrote Irving.

The lust for sucrose seems to be universal, common to all mammals, as anyone who has seen a dog destroy a box of chocolates will know (and then had to pay the vet's bill). We do not know when beekeeping, rather than bee-raiding, became a skill but it would have been a prized one: honey, transportable, long-lived, was one of the early trading goods.

Sweetness enters Western culture and language early as a metaphor – music is 'honey-sweet' in Homer's *Odyssey*, while Moses in the Bible's Exodus promises the Israelites a 'land flowing with

118

milk and honey' – but a few hundred years later, around 1000 BCE, a much more easily obtained sweetness, the refined juice of the sugar cane, began to emerge from northern India.

> Sugar came from north India via the Middle East: the words Europeans use for it trace that journey. The Arabic *sukar* is an adoption of the Sanskrit (northern Indian) *sharkara*, originally meaning small granules; 'candy' comes from the Arabic/Sanskrit for sugar itself, *khandakha*.

SUGAR CONQUERS THE WORLD

The Indian crystals – originally great lumps of cane sugar that had to be shaved or chipped – were to change the course of history for many disparate peoples. But they did not catch on fast: cane sugar was gradually traded further west and east over the next millennium. Islam brought it to North Africa and then Europe. One thousand years ago, sugar cane was being cultivated on the Mediterranean islands and in Spain. The English and Scots may first have encountered the new concentrated sweetness when fighting in Palestine during the first Crusade of 1095. Like everyone else, they were entranced; they brought it home.

Household accounts from the Middle Ages tell us about 'Marrokes', 'Cypre' (Cyprus) or 'Babylon' – names for the strange, sweet lump of crystals borrowed from their place of origin. Initially used in medicine, sugar soon became a feature of the store cupboards of aristocrats' houses. King Edward I's household accounts record the purchase of 1.3 tons of sugar in the year 1287, some of it mixed with violet and rose petals.

By the late Middle Ages, recipe books show sugar being used with the joyous abandon of today – in stuffing for game, for veal roasts and liver. Meanwhile, in the cookery of the royal court of Siam, sugar was being used to balance sour, salt and bitter in precisely measured piquant sauces that became and remain the nation's pride.

SUGAR ARCHITECTURE

Sugar's most spectacular change comes at a temperature a little above water's boiling point. Its molecules separate and then reconnect themselves end-to-end in crystals as they cool, giving us superfine, rigid columns; from these come spun sugar, the mirror shine on a toffee apple, glazes and butterscotch. These have thrilled creative humans: castles, ships, entire gardens and even erotic tableaux recreated out of sugar have entertained the wealthy at banquets since the Renaissance.

Architectural fantasies made of spun and hardened sugar as table decorations seem to have been de rigueur for some 300 years. King Henry VIII employed seven cooks for a sugar banquet in 1526: they built a dungeon, a manor house with swans, a tower and a chess board – the sugar 'burnished with fine gold'. There is a sad picture of Christina, Queen of Sweden, at a state dinner with Pope Clement in 1655; the poor woman, seated alone, yards away from her host, is watched by a crowd of courtiers. She is just half visible behind the ranks of baroque sugar 'subtleties', made by Italian sugar chefs, then known as the greatest.

Practising the sugar arts before thermometers was not easy. Early cookbooks helped by explaining what sugar would do in different states and at different temperatures but lots of burnt fingers were likely involved. *Le Confiturier François*, published in 1655, suggests dropping molten sugar between finger and thumb to test it. One key stage is reached if the drop doesn't flow off and remains 'round as a pea', a second when 'on opening the fingers, it forms a small thread'. Then the molten sugar is ready for using in jams and preserves. Still hotter, it forms shapes 'like dry feathers without stickiness'. This sounds like the 'hard cracking point' that occurs above 149°C.

Marmalade-makers know they need to get the molten sugar to the thread or wrinkle point: that is, at 105°C and involves boiling away all residual water. Once we would drip the liquid sugar onto a cold plate and look for the wrinkle as it began to set – now only die-hard traditionalists will not reach for a thermometer.

One of the last great sugar feasts was given by the Prince Regent, the future King George IV, at the Brighton Pavilion on 15 January 1817. In charge was the first celebrity chef, Antonin Carême; after the first course he sent in set pieces constructed of sugar: 'The Chinese Hermitage' and 'The Ruin of the Turkish Mosque'.

As ever, these were designed to be admired rather than eaten: their modern echo is in the tiered glories of a traditional wedding cake. Nonetheless, *The Food Programme* reported on a modern event for sugar artists, the biennial Sugarcraft, held not far from the Brighton Pavilion. 'We've got a mosaic picture by Klimt, a chocolate clock ... We've a chocolate cake here with a bas-relief cowboy,' said one sugar sculptor. 'The magic of sugar! You can eat your mistakes as well!'

POSH TEETH

By 1817, the Prince Regent's feast making gloriously wasteful fantasies out of sugar was a bit old hat – sugar, once a luxury so expensive that stores of it were kept under lock and key, had become cheap enough to be a staple in most British households. Our consumption of it had gone from some 1.8 kilos each a year in 1710 to over 8 kilos by the end of the century. Everyone now had access to instant sweetness, but at a cost.

As sugar took its crucial role among the wealthy and privileged of Europe, an unforeseen side effect occurred: the rich stopped smiling. The aristocrats of the French and English courts were notorious for their dental problems: ambassadors reported Queen Elizabeth I only had a few, discoloured teeth left. In portraits

'Her nose is a little hooked, her lips narrow, and her teeth black, a defect the English seem subject to from their great use of sugar.'

Paul Hentzer, a German traveller who visited Elizabeth I's court in 1598

> **Sugar straight-up**
>
> Nowadays, we disapprove of eating sugar by itself, for itself –
> except of course in sweets. But it wasn't always thus. One of the
> treats of holidays with my Scottish grandmother was a 'sugar
> piece': a sandwich of two slices of Mother's Pride bread buttered
> with granulated white sugar poured between them. The crunch
> was the best bit.
>
> We'd eat Tate & Lyle's golden syrup the same way and drink a
> lot of Barr's Cream of Soda too, which was lemonade without
> lemon in it – just soda water and sugar. In Scotland's industrial
> age, this diet was something people took pride in: it was said
> that the Clyde shipyards were fuelled by sugar – 'a jilly piece [jam
> sandwich] and a lemonade' being a worker's normal midday
> meal. The other side of this nostalgia is that Scots remain some
> of the least healthy, most overweight people in Europe.

from that time onwards, nobles begin to keep their mouths firmly closed. Archaeological research in old burial grounds reveals that the teeth of the rich were in very bad condition, far worse than those of ordinary people.

The first physician to make the connection between sugar and dental decay worked for Louis XIV of France. Pierre Pomet used his position to publish a 'history of drugs' in 1695 – it was translated into German and English. He wrote five pages on sugar and its uses, culinary and medicinal; it was, apparently, good for the lungs, the kidneys and the bladder. But Pomet warned 'it rots and decays the teeth': his client King Louis lost all his teeth before he was forty. According to the historian James Walvin, in the European courts, smiling or laughing became unacceptable: a sign of vulgarity.

Nonetheless, by the mid-seventeenth century, sugar's democratisation was underway. It was available in shops in provincial British towns, both as a medicine for ailments from consumption to sores and as an ingredient in cooking and preserving. It has been argued that this may have actually improved ordinary people's diets. A whole host of bitter fruits, including redcurrants, damsons and sour apples, now appear in recipes, with sugar to take the acid edge off. As with salt, cocoa and so many other once rare, foreign luxuries, demand from aspirational northern Europeans stimulated entrepreneurs. More, cheaper sugar was a way to make a fortune.

SUGAR AND AN AFRICAN HOLOCAUST

Sugar cane only grows in the tropics, with copious water. The cultivation and then the crushing, boiling and refining of the sweet cane juice is labour intensive. What was needed for the early seventeenth-century sugar entrepreneurs was land in hot, wet countries nearer Europe than Asia and, of course, a lot of cheap human labour.

These were both available around the Atlantic. Improved ship design and more precise navigation techniques were now making trans-oceanic trading voyages less of a gamble. Another system, that we'd now call private equity or venture capitalism, was gearing up to finance these expensive expeditions. The potential of the new land was obvious: Christopher Columbus brought the first sugar cane plants to the Caribbean island he called Hispaniola on his second exploratory voyage in 1493.

So, a little more than a century after Europeans had 'discovered' the New World and begun raiding its mineral wealth and enslaving the native peoples, the land was put to use to make money in a new, industrialised way. Tobacco-growing for export was already lucrative by the 1620s in what would be the American state of Virginia. But it was the introduction of sugar cane to the British colony of Barbados that would come to change the entire Caribbean, exterminate its native populations and devastate the civilisations of West Africa.

In the 1630s, three English settler-businessmen in Barbados, James Drax, William Hilliard and James Holdip, decided to give up growing tobacco on their land: prices were being undercut by better quality crops from Virginia. So Drax travelled to Brazil to look at Portuguese sugar cane planting. When he returned, the men began growing cane and experimenting with the complicated boiling and refining of the cane juice. It took ten years but eventually they got it right. By 1650, the island was exporting 7,000 tons of sugar a year back to Britain.

The other necessary tool for the work was already in place: enslaved human beings were being shipped over from West Africa to work in the plantations of the Americas. A monopoly set up by the English King Charles II and his brother, the future king James II, was the main transatlantic trader in enslaved people until 1691. (The Royal Africa Company stayed in business until 1731, at which point the monarchy and its merchant co-owners had shipped over 200,000 enslaved people.) As the eighteenth century wore on, demand for sugar in Europe increased and more and more land in the Caribbean was turned over to sugar cane.

That caused the numbers of Africans required to grow and make the sugar to soar, and a hugely lucrative trade boomed. Manufacturers of all the things necessary for trading sugar and people – from shipbuilders to gunmakers – profited, and so did the British government as it taxed everything from enslaved people to sugar itself.

In 1746, the economist Malachy Postlethwayt pointed out the obvious: Britain's sugar addiction had created a slavery-based industrial complex too huge to tamper with. Without enslaved people no sugar, rum or tobacco and 'consequently the public revenue, arising from the importation of plantation produce, will be wiped out. And hundreds of thousands of Britons making goods for the triangular trade [between Britain, West Africa and the Caribbean] will lose their jobs and go a begging.'

By the mid-century there were 65,000 enslaved people on Barbados – nearly four times the white population. Barbados was by then Britain's most successful sugar colony, and one of its most murderous. On most of the Caribbean islands, the average enslaved adult working on the sugar plantations survived four years after arrival: on Barbados, it was only three – perhaps accounted for in part by the fact that punishments there were notoriously harsh. As the numbers soared, Drax and the other planters learnt to hate and fear black people.

Sugar planting in Jamaica, Antigua, Grenada, Guyana, Saint-Domingue (Haiti), the Virgin Islands and other Caribbean colonies of British and European countries boomed, too: by 1800, there were more than 300,000 enslaved people in British Jamaica, most of them involved in sugar making. The profits from the sugar were so great that the price of an adult African could be recouped in just over a year: it was cheaper to work the enslaved people to death than to look after them.

No one knows how many Africans died enslaved in the 250 years before abolition in the British colonies. It was a genocide: we know that 3.25 million African people were shipped by the British to the Americas but their children, all automatically enslaved at birth, are uncounted. When slavery was abolished in the British Caribbean in 1833, only 665,000 enslaved people had survived, fewer than were there in 1807.

On Barbados, Drax, his partners and their heirs became immensely rich. His direct descendant, the British MP Richard Drax, still lives on the huge estate in Dorset paid for by sugar and slavery, when not holidaying in Barbados at the 'Great House' whence the plantation was run. Other fortune seekers took notice and so did the European colonialising powers. In 1667, the Dutch swapped an obscure but strategic island on the east coast of North America – Manhattan – with the British for several tropical territories, including the sugar-producing colony of Surinam on the Caribbean coast of South America.

By 1770, France's half of Hispaniola, the colony of Saint-Domingue – soon to be renamed Haiti – was the world's biggest sugar producer, sending 60,000 tons a year to Europe. British Jamaica, a hundred miles west, then exported only 36,000 tons but production was increasing every year. In France, the result was a craze for sugared coffee: a British visitor wrote that the French used so much that your spoon stood up in your cup. The results, in diseased gums and faces hollowed by missing teeth, were as if, Walvin writes, 'the enslaved were getting their revenge for the abominations heaped upon them in the Caribbean'.

'I will never more drink Sugar in my Tea, for it is nothing but Negroe's Blood.'

Royal Navy purser Aaron Thomas, writing after his ship visited the West Indies in the late eighteenth century

In August 1791, the enslaved people of Saint-Domingue rose up against the French sugar planters; they soon had to fight both them and the British, who saw a chance of seizing the island for the sugar machine. Forty-thousand British troops died and the republic of Haiti survived – the result of the only successful rebellion of the enslaved in the whole era.

The fifty-year battle to abolish the British slave trade and then slavery itself was a popular movement. From its beginnings in the 1780s, it was led by ordinary Britons and a handful of escaped or freed people of African origin. Women were to the fore, leading a campaign that from the beginning targeted slavery's most obvious result: sugar. They took the fight into middle-class homes, to the tea table.

Pamphlets from the abolitionist campaign warned people that sugar was a 'blood-sweetn'd beverage'. One claimed that entire 'roasted' Black people had been found in sugar barrels on arrival in Britain. Petitions were signed and mass boycotts of West Indian sugar began.

One of the reasons it proved so hard to abolish slavery was the British states' obvious dependence on its massive revenues from taxing sugar and all the other things necessary to plant, process and ship it. It was a reliable cash flow that underpinned the borrowing to finance long wars against the French at the turn of the eighteenth and nineteenth centuries. Britain's rising sugar addiction was – as with tobacco and alcohol – a boon to the British Treasury.

But by the 1840s most sugar plantations on Jamaica and the other British islands had closed. Within ten years of the end of British slavery, the Victorians were already painting the British as the heroes of the slavery story. The country had led the world in abolishing its practice, so the narrative went, though the truth was that Britain had made more money out of it than any other country in Europe. The irony of this self-satisfied and dishonest story is that not only did the British leave their former slaves in dire poverty but they continued to buy sugar made by slaves. It was cheaper from Cuba and Brazil – from which companies like Lyle's were still shipping it until slavery ended there in the 1880s.

Sugar was the reason for most British enslavement of Africans and it is hard to conclude that the story has done anything but blight millions of lives, right up to today. It was, as the historian Simon Schama puts it, 'the original sin' of the British Empire.

Ordinary Britons' complicity in transatlantic slavery, and some people's continuing denial of the legacies of it, is a matter that still troubles us today. Speaking on *The Food Programme's* discussion of food and the legacy of slavery in August 2020, the American food historian Dr Jessica B. Harris discussed the eighteenth-century poet William Cowper's 'Pity for Poor Africans', which was written in support of the abolitionist movement. In the run-up to the parliamentary votes on the slave trade in the 1790s, thousands of copies of the poem were printed and distributed under the title 'A subject for Conversation at the Tea Table'.

> *I am shock'd at the purchase of slaves,*
> *And fear those who buy them and sell them are knaves;*
> *What I hear of their hardships, their tortures, and groans,*
> *Is almost enough to draw pity from stones.*
> *I pity them greatly, but I must be mum,*
> *For how could we do without sugar and rum?*
> Excerpt from William Cowper, 'Pity for Poor Africans', 1780s

130

Cowper was highlighting the hypocrisy of the average Briton, who, well aware of the horrors of the slave trade, would tolerate it because of the benefits for eighteenth-century Britain in cheap sugar, rum and, of course, the immense wealth generated for planters, shippers and sugar merchants. 'He has a lot of excuses,' said Professor Harris of Cowper and his morally conflicted sugar-user at the tea-table. 'The last line of the poem is "He shared in the plunder, but pitied the man".'

HIGH FRUCTOSE AND BAD FAITH

It is now more than one hundred years since doctors started recommending people cut down on sugar or replace it with something healthier. While the link had already been made between bad teeth and sugar, it was during the famine in Paris during the Prussian siege of the city in 1871 that physicians spotted lowered glucose levels in the urine of diabetics. Paris was starving: restaurants served cats, dogs and rats, and the Paris zoo's only two elephants were slaughtered for meat. Doctors speculated that the decrease in glucose levels was due to the lack of carbohydrates, including sugar, in the Parisians' diet. Based on this discovery, suggested treatments for diabetes began to mention low carbohydrate diets from the 1870s.

Sugar's relationship with weight was even less understood then and the causal link between sugar and type 2 diabetes is still only hypothesis today. But it seems clear that the overall increase in average weight in the populations of many richer countries has been concurrent with increased sugar consumption. We only have reliable mass data covering the last fifty years but in that period the number of Americans who are obese has gone from fourteen to forty per cent. Twenty-eight per cent of adult Britons are obese: twice as many as in 1993–5.

SUGAR'S REBOOTS

Since the nineteenth century, manufacturers have worked at delivering sugar in ever cheaper and more user-friendly forms. In 1885, the Scottish businessman Abram Lyle seized on a way of turning the molasses left by sugar refining, then used as pig feed, into something more valuable. Tate & Lyle's golden syrup was the product, sold on the fact that it would always stay liquid. That is true: Captain Scott took tins of it on his ill-fated expedition to the South Pole in 1911. When some were found among his abandoned stores in Antarctica in 1965 the tin and syrup were still in usable condition.

The design for the green and gold tin, first used in 1904, is now Britain's oldest essentially unchanged food packaging. The picture on it of a dead lion represents a biblical riddle, told by the hero Samson after he found bees nesting in a lion's carcass: 'Out of the eater came forth meat and out of the strong came forth sweetness,' he declared.

The most notorious re-engineering of sugar is a substance made from the starches in corn or potatoes and is known as HFCS – high fructose corn syrup. The hook here is using more fructose, the sweeter of the two molecules that make up sucrose (ordinary table sugar). It is about twice as sweet to the taste receptors as sucrose but it's hard to make stable in crystal form.

Sweet syrups, mainly of glucose, had been made from corn and barley since 1811, by heating them with sulphuric acid. Treating those with *Aspergillus niger* – from the same family of bacteria that is used in processing soya – turns some of the glucose to fructose, in a viscous syrup that makes food chewy and is acid enough to react in the leavening process in baking.

HFCS was easy to manipulate and cheap, so from the 1980s manufacturers started to use it instead of sucrose in many foods, especially sweet fizzy drinks like colas. Now, the United States consumes more corn syrup than sugar produced from sugarcane or sugar beet, the traditional sources. HFCS is actually only five per cent higher in fructose than table sugar, but since the public discovered it, campaigns against it on health grounds have become common.

There is not in fact much solid evidence that fructose is more damaging than the other forms of ordinary sugar; it's the increase in consumption brought by HFCS's cheapness that is the problem.

Glucose has also been linked with a risk of heart disease. The body metabolises different sugars in different ways, but these pathways are still not completely understood.

Much sugar science is still hotly contested. Dr Robert Lustig – he of *Fat Chance* – leads what is emerging as a current scientific consensus: sugar's interference with the liver and normal insulin production is the key problem. It is this that is thought to account for the extraordinary rise in the incidence of type 2 diabetes that used to be called, because it was rare, 'early onset'.

Radical reduction in added sugar consumption is his solution: 'There is not one biochemical reaction in your body, not one, that requires dietary fructose, not one that requires sugar. Dietary [added] sugar is completely irrelevant to life. People say, oh, you need sugar to live. Garbage.' Another much-debated problem in the rich world is an equally shocking rise in cancer of the lower digestive system: there are theories that the rise in our sugar consumption is linked to that too.

What is clear is that what you eat the sugar with is important. 'Eat the apple, not the juice', goes one piece of advice – the point being that you will digest the sugar more healthily if it comes with the fruit's fibre. Also, of course, if you eat an apple you will feel fuller in your stomach – what's called satiety. Drinking an apple juice – or a cola – will leave you still feeling hungry.

Despite all the debates and the health claims, the fact remains that sugar is sugar: a simple chemical. The fact that the sucrose of table sugar is refined does not mean much – it is not different in any significant way from 'natural' sugar in fruits, it is just that our bodies metabolise the two differently. Labels that promise 'only natural sugars' or 'fruit sugar' are offering little if that sugar has

133

been separated from the fruit it started with. They may well be covering up another assault on our livers and our teeth.

SUGARALIKE AND OTHER CHEMICALS

In an attempt to lower sugar in our diets, a host of substitutes have emerged since the 1970s, most of them backed by food corporations keen to make sure any revenue from sweetening – natural or synthetic – stayed in their hands. One of the most successful new sugars, aspartame, became the market leader after it got regulatory approval in the United States. It made the reputation and the fortune of the company, Searle, and its chief executive, Donald Rumsfeld. His brilliance in business led to a job as George W. Bush's defence secretary, and chief architect of America's post 9/11 War on Terror.

Many questions remain over the long-term side effects of artificial sweeteners. Some have been shown to cause cancer in laboratory animals – but at far higher doses than humans consume them. Some may trigger the same insulin release that glucose does. There is also research that implicates artificial sweeteners with alterations to gut bacteria and a range of consequent problems in the liver, lymph and spleen.

Few sweeteners have any natural origin – no matter the claims on the labels. Even the stevia plant's sugar is bolstered for consumers' use with other chemically derived sugars. Saccharin, twice as sweet as sugar, is made from coal tar. Saccharin was already available in the nineteenth century as a sugar substitute, prescribed to diabetic people. In the 1950s, it was combined with a chemically derived sugar, cyclamate, and marketed with great success as 'Sweet'N

Low' – even though American manufacturers were well aware of the drug's carcinogenic potential. Cyclamate was – after a long legal fight – banned in the USA in 1969, though it is still used in Europe. (Sweet'N'Lo now uses saccharin, dextrose and tartaric acid, which are considered safe.)

Sweeteners haven't fully caught on with consumers, mainly because of issues of side-tastes, and also, crucially, the fact that no artificial sweetener actually works like sugar when you cook with it. The substitute sugars' greatest success has come where there are other tastes to mask them – as in highly processed 'low-sugar' jams and sweets and in soft drinks. They also have been a boon for people who cannot eat much sugar for medical reasons, such as those with diabetes.

DIRTY FIGHTING

The other story of sugar's journey to demonhood is of a furious fight-back from the industry – the heirs of the sugar planters and merchants who fought so hard to stop slavery being abolished. As with other contentious foods in this book, these battles have often been contested most aggressively over the work of scientific researchers who have questioned the safety of highly profitable ingredients.

In 1972, the dietary scientist Professor John Yudkin rang the alarm bell over sugar. He made public physiologists' and health researchers' suspicions about the damage being done to the body's systems by the great rise in sugar consumption. 'If only a small fraction of what we know about the effects of sugar were to be revealed in relation to any other material used as a food additive,' wrote Professor Yudkin, 'that material would promptly be banned.'

When he published his book, *Pure White and Deadly*, in 1972, Yudkin was Britain's leading nutritionist. His career was all but destroyed by the sugar industry, abetted by complacent nutrition scientists.

Rather than starting a necessary discussion around sugar and public health, Yudkin's work was dismissed. 'An English biochemist described my father's work as nothing more than scientific fraud,' his son Michael told *The Food Programme*. It was not. He specified obesity, heart disease, diabetes and tooth decay as the problems associated with excessive sugar, all of which are now agreed. John Yudkin died, largely forgotten, in 1995, but his pioneering book has now been republished with a foreword by Dr Robert Lustig. 'Holy crap!' Lustig thought when he found Yudkin's work. 'This guy got here thirty-five years before me.' If Yudkin had been given a fair hearing, an unknowable number of early deaths might have been avoided.

The campaign by the sugar industry, its lobbyists, scientists and bought politicians has been fierce and dirty. It is very similar to the one waged in the 1960s to stifle scientific findings about tobacco smoking and evade government controls. One tactic was to blame saturated fat – another contemporary demon – for the sins of sugar. In 2016, a scientific review in the United States uncovered many instances over fifty years in which Coca-Cola and other sugar-dependent corporations paid for supposedly independent research to back their lobbying.

'They were able to derail the discussion about sugar for decades,' says Stanton Glantz, a professor of medicine who conducted the investigation into academic archives. One example found was a large payment in 1967 by a sugar industry foundation to three Harvard scientists, who were due to publish an important review of research on sugar, fat and heart disease.

Dr Glantz told the *New York Times*: 'The studies used in the review were handpicked by the sugar group, and the article, which was published in the prestigious *The New England Journal of Medicine*, minimised the link between sugar and heart health and cast aspersions on the role of saturated fat.' During their research, letters show, the scientists made sure to consult with their industry paymasters to make sure their findings were approved. 'By today's standards,' says Glantz, 'they behaved very badly.'

A BETTER RELATIONSHIP WITH SUGAR?

There are signs that our 300-year-long sugar party may be entering its late stage. No one – not even the sugar industry – can any longer challenge the science on sugar and health. The damage done by excess sugar in the diet is visible in every hospital and every school in the developed world. Thus, with every few years, health guidelines get stricter: since 2015, the World Health Organization has recommended that no more than ten per cent of energy intake should be from sugar; for an adult on a 2,000-calories-a-day average diet that translates as 25–30g, or six-to-seven teaspoons of sugar a day – less than is in a generous Nutella sandwich or a can of cola. In Britain, the recommendation is for no more than five per cent of energy intake to come from 'free sugars' – excluding those naturally occurring in fruit (though not fruit juice) and vegetables. It puts the upper limit for an adult at 30g a day.

That's all very well. Consumers get lectured too much nowadays and health guidelines that interfere with our pleasures are the most ignored. Both the National Health Service and the World Health Organization have revised theirs, issuing edicts asking people to eat less and less sugar in recent years. In the same time, per capita sugar consumption has, overall, gone up in the UK. People may think they eat less sugar: health-conscious consumers often report a drop in their personal consumption. But that ignores the fact that much of the sugar they eat is, like salt, hidden in manufactured and processed foods.

For the first time, though, these dietary recommendations are being weaponised. Governments can see that it worked with tobacco. Punitive taxes on high-sugar drinks have been introduced in several countries, including the UK, despite furious fight-back from sugar-industry-funded lobby groups. In Mexico and Scotland, two of the world's most overweight nations, sugar taxes are already being hailed a success. They are, little by little, bringing consumption of sugary drinks down.

This has something to do with greater publicity around sugar's impact on health. But more significant has been the reaction from manufacturers. The sugary drinks market is cut-throat and

Want to reduce your sugar?

Tests have shown that in many traditional biscuit and cake recipes – ones where sugar is not crucial to the structure of the food – the amount can be reduced by twenty or thirty per cent without a noticeable problem with taste. But, of course, the easiest way to cut down on sugar in your diet is to reduce your consumption of processed foods.

very sensitive to price differences. When the British government introduced its small levy on such drinks in 2018, it found that half of all manufacturers had reduced sugar content even before the new law took effect, in order to avoid having to raise prices. Early in 2022, the British government began to consider the highly controversial suggestion, made in its own National Food Strategy, that all sugar going into processed food should be taxed.

The response to the lobbyists who opposed sugar taxes as 'a tax on the poor' – there were no price rises. Meanwhile, tougher labelling is tackling high-sugar foods people thought were healthy: fruit juices, muesli and nutrition bars.

Surveys now show that a once fat-obsessed British public is more concerned about sugar in its diet. Nestlé and Mars Wrigley, two of the international sugar giants, are trying to get ahead of that interesting trend by announcing voluntary future reduction targets in their overall use of sugar, as they did a decade ago with salt and trans fats.

There may also be technological solutions, though these are usually hard to sell to a public that has become bored and distrustful of food bio-tech. In late 2020, DouxMatok, an Israeli start-up, announced that it was taking an extraordinary invention into production. It was a sucrose that tastes forty per cent sweeter; they say it will allow manufacturers to reduce the sugar in their recipes by the same amount. This seemingly impossible act was not done through manipulating levels of glucose and fructose, as HFCS does, but addressing the human end of the process: an inefficiency in the taste receptors on the tongue.

Avraham Baniel, an Israeli scientist who patented the new sweeter sugar when he was ninety-six years old, had noted many years earlier that adding starch to sugar made it taste sweeter. This is because in a bite of a sugar-laden snack bar, for example, only about twenty per cent of the sugar is actually tasted. Just one in five sucrose molecules meet a receptor in the mouth. The rest just goes straight into our bellies – needlessly adding to our sugar consumption and excess calorie intake. Baniel found that by bonding the sugar crystals to inert silica he could make more of them reach the taste receptors. The analogy used is candy-floss – where the re-engineering of the sugar into fine strands spreads it further and makes the treat much more sweet than a similar weight of granulated sugar.

The prospect of a less harmful sugar is so alluring that many companies are investing fortunes in it. Some are seeking out new sugars in the natural world – there are endless ways in which the atoms of carbon, hydrogen and oxygen that make up sugar can be arranged, and each one acts slightly differently on our taste receptors and in our gut. New sugars have now been isolated from algae, aphids, shellfish and mushrooms.

Tate & Lyle, who brought us the questionable benefits of the sugar cube and golden syrup, are marketing a corn-derived synthetic based on a rare natural sugar, allulose, that occurs in figs and maple syrup. The company claims it is seventy per cent as sweet as sucrose and tastes quite similar. But because it passes through the body largely undigested it only has ten per cent of the calories and the company claims it does not raise blood glucose or insulin levels. As important: you can cook with it. As a Tate & Lyle spokesperson told *The New Yorker* magazine, 'It behaves like a sugar because it is one.'

The company has for forty years had a brand-leading sugar substitute, Splenda, and in 2015 it began selling allulose under that name and that of Dolcia Prima in the United States. The sell to consumers was that it was a plant-derived sugar – it mentioned figs – and so it was natural, it didn't have to be called sugar. Impressed, business analysts proclaimed that it might be 'huge' enough to save the ailing old sugar-dealing company.

Yet, seven years after its launch, allulose has yet to make much of an impression on the US market. It is only available in health shops in the UK and at nearly ten times the price of granulated white. Regulatory approval in Europe has been held up by laws against genetic modification.

Scared of sugar though we are today, many food manufacturers are dealing with the threats from government by simply reducing it in products in the hope that consumers won't notice. A similar tactic has worked with salt and it remains a truth in cooking, as in much else in life, that you can reduce consumption of something by five or ten per cent quite easily. It is the next bit that is tough.

It may be that the century-old effort by health experts to get us to reduce our sugar consumption to something a little closer to the pre-eighteenth century, pre-Caribbean slavery norm, will be won not by technology but by the simplest means, price. It is always the strongest incentive. Because, in the words of *The New Yorker* magazine writer Nicola Twilley, summing up her 2020 account of 'the race to redesign sugar': 'Just as the only good substitute for sugar is sugar, the only good way to eat less of it is, sadly, to eat less of it.'

CHAPTER 6

APPLE OF THE EARTH

POTATO

'Potato growing enabled grain to become more of an
economic, negotiable item. As the industrial revolution
got going, people were able to grow potatoes, live
off potatoes – the best single bundle of nutrition
there is. It fuelled population growth wherever
it arrived, excess population found its
way into the cities, cities were the
basis of factories, the potatoes
fed the labour, and that got
the cycle going.'

Historian John
Reader, *The Food
Programme*,
2009

Donald Brown is a crofter on the Hebridean island of Tiree. Like many in northern Europe's poorer places, as a child the potato was central to his nutrition, a key to life. Growing up in the 1940s on the eight-mile-long island off the west coast of Scotland, he ate them every day for dinner at two o'clock – the day's main meal.

Donald's mother would boil the *bhuntata* – as they are called in Gaelic – unpeeled, in sea water: much better than water from the well, he says. 'She would put them on the table, whole, with a bowl of home-made butter, buttermilk and soft cheese.' Sometimes there would be a little beef to go with the potatoes, if the family had slaughtered one of the dairy herd. As a treat, Donald's mother would sometimes fry the potatoes in beef dripping. But young Donald's favourite meal was *bhuntata* and *sgadain*, potatoes with fried herring, caught in nearby Gott Bay in October and November.

The family grew potatoes on the croft. As a child, Donald helped tend them; a job for most of the island's children. He remembers long hours spent chopping potatoes as emergency winter feed for the cattle. In October, all the Tiree children got two weeks off school to help with the potato harvest. Scottish children still have eighteen-day half-terms in October: the 'tattie holidays'. But very few spend the time out with a grape, the traditional three-pronged fork, picking potatoes.

The tubers were a hard fact of island life, as important and omnipresent as the Bible and the wind. Donald finds it hard to imagine a world without them. He remembers an old island story, much quoted: 'There was a wee lad who was asked by a tourist what he ate at home. "Mashed potatoes for breakfast, mashed potatoes for dinner and mashed potatoes for tea." The visitor asked, "What does your Mummy give you with them?" The boy answered, "A spoon!"'

The Brown family grew up well aware that dependence on potatoes was a dangerous thing. Just one hundred years earlier, a series of failures of the crop because of disease had starved and killed thousands of people of the islands and Scottish Highlands. In Ireland, the death toll from what became known as the Great Famine was far higher, amounting to perhaps one million people between 1845 and 1852. The potato failure combined with endemic poverty forced many in Ireland and Scotland to emigrate; Donald is still in touch with Brown cousins, descendants of family that went to Canada in the nineteenth century. The potato, or lack of it, reshaped the North Atlantic world. Like millions of others from Britain and Ireland's poorest rural fringes, their migration was forced by the combined pressures of hunger and landlords determined to make more money from the land.

For Donald Brown, this is not ancient history but recent and tangible. Why his family and so many others in rural communities

'Those who are habitually and entirely fed on potatoes live upon the extreme verge of human subsistence, and when they are deprived of their accustomed food there is nothing cheaper to which they can resort. They have already reached the lowest point of the descending scale, and there is nothing but starvation and beggary.'

Official report on poverty and famine in Ireland and Highland Scotland, 1848 came to be so dependent on the potato has important lessons as we think about the staple foods we rely on in the twenty-first century.

A STRANGE ENCOUNTER

That story begins 1532, in Peru, when Spanish explorers led by Francisco Pizarro first met the potato. It would have been a supremely alien object to a culture that then knew nothing of tubers – yams, potatoes and sweet potatoes – as food. But Pizarro and the *conquistadores* ate the root and recognised its value as an easily stored, easily digested provider of concentrated nutrition and – though they did not know it then – a source of vitamin C. The potato soon made the journey to Spain and then to the rest of Western Europe.

The earliest potatoes' defences made people sick. The Andeans had to eat them with clay, which absorbed the glycoalkaloid toxins. Some ancient human had spotted wild animals licking the earth and then eating the potatoes, and copied them. Potatoes without the toxins were not bred until about 2,000 years ago; in the Andes you can still buy the ancient varieties, along with a little packet of earth to chew with them. Potato historian Charles Mann told *The Food Programme* he'd tried this: 'it's quite tasty'.

Mann was less keen on some of the other Andean ways of serving potatoes: *tocosh,* a stinky fermented potato pulp dish that has defeated the stomachs of most visitors to the Andes is one example. But he admires the extraordinary industry involved in the development of the tuber. 'The Andean peoples bred potatoes for every conceivable type of conditions you can imagine – wet areas, dry areas, cold areas, warm areas. As a result, there are thousands of different varieties of potato: the International Potato

Center in Lima has more than 7,000 named varieties.' Hence the potato's ubiquity.

Potatoes are seventy per cent water and so they go off easily. They can become green and poisonous if exposed to sunlight. The Peruvians worked out ways of preserving them, by drying, smoking and fermenting them. *Chuños,* a freeze-dried and sometimes smoked potato that keeps for years and looks like small grey marshmallows, has its fans. The Peruvian chef Virgilio Martínez serves them in his London restaurants with cheese sauce and peppery leaves.

POLITICS AND THE POTATO

Like all the other arrivals from the Americas – except tobacco – the potato took time to catch on. As so often with the new foods, European difficulties began with the categorising of them. Like the taxonomising of the tomato, the most easily comparable existing food item was the apple. It kind of looked like a potato. So, in French and Dutch, potatoes are 'apples of the earth'.

German uses a variant of the word for truffle to name potatoes *kartoffel*. Other languages adapted the Quechua name, *batata*, which has survived in Gaelic. But, of course, the new food could not be eaten raw and did not taste remotely like an apple or a truffle. Potatoes lacked good branding and take-up was slow.

But they caught on. The first Northern Europeans to cultivate potatoes as more than just a curio may well have been the Irish. The pleasing folk explanation for this is that refugee crew of the Spanish Armada ships, wrecked on the Irish coast after the great battle of 1588 in the English Channel, brought them ashore as gifts

The wrong name

The word of the main Andean people, the Quechua, for potato is *papa* – and that's now used throughout Latin America and in southern Spain. The Quechua for sweet potatoes, a very different tuber with a papery skin and orange flesh, is *batata*. When the Spanish adventurers took the two tubers back across the Atlantic, they knew the difference. But somehow on the journey further north and east across Europe, the names were confused, for ever.

for their rescuers. Spanish traders bringing them from the Canary Islands is a more likely explanation.

A definite hindrance to their acceptance in Christian Europe was the fact that the Bible has no mention of potatoes. Some theologians deduced they makes a food for heathens or animals only. But the ease with which potatoes grew and flourished eventually outweighed these concerns. It is an amazing performer: in ideal conditions a field will grow four times as much food in potato form, as it will a grain crop like wheat or barley. Besides, by the late eighteenth century, the potato was fashionable, not least at the court of Louis XVI of France, where the taste for a season demanded that men and women wore a purple potato blossom in their wigs.

Around the same time, reforming rulers like Frederick the Great of Prussia and Catherine the Great of Russia started promoting the potato as an answer to poor harvests and famine. They ordered their – often unwilling – subjects to cultivate and eat them. Tsars after Catherine continued the potato crusade. In 1840, Russians rioted in several cities against an order to plant common lands with potatoes.

They were not alone. The potato had become a political object. Radical campaigners in Britain condemned it as a symbol of the oppression of the poor – and the latter agreed. 'We will not live on potatoes,' read a protest banner raised by agricultural labourers in Kent in 1830, during a series of riots in south-east England over lack of land and poverty. For the followers of Karl Marx and Friedrich Engels, writing the following decade, the potato was a symbol of oppression.

Nonetheless, potatoes were now rivalling bread as a staple carbohydrate across much of Europe, doubling the food supply and, with an alternative to cereals, making disaster less frequent. That led to population growth, increased incomes for smaller farmers and resulting social change. 'Without the potato, no European empires,' says Charles Mann. Across the continent, regimes that had been regularly destabilised by recurring famines were able to increase their wealth and power. Modernity was clearly potato shaped, though by the mid-nineteenth century the early nutritionists were worrying that potatoes made workers stupid and sluggish, and were the cause of ever-rising immorality, a perennial Victorian concern. It is not surprising that dependence on them did not translate into much affection for the potato until later in the century. A key was the happy marriage of cut potato and very hot fat: chips, or French fries, finally arrived in British cities in the 1860s, probably from Belgium.

People have fixed notions about potatoes and every potato cook has their favourite for particular purposes: floury, waxy, new, old. Many potato varieties don't look dissimilar, yet we will pay very different amounts for different species. As a result, potato fraud is big business. The much desired Jersey Royals have had to be given Protected Designation of Origin status across Europe, like Puy lentils and Parma ham.

In 2008, the British Food Standards Agency tested 294 potato samples and found that thirty-three per cent were not what they claimed on the label. Many called King Edward – a much-loved, multi-purpose variety that celebrates the coronation of King Edward VII in 1902 – were in fact Ambos, a standard big farm variety that is resistant to blight and sells for half the price.

FAMINE AND A RESHAPING OF THE WORLD

The people of the west of Ireland, early adopters of the potato, were encouraged to grow more by everyone with power over them. Landowners who liked the efficiency of the potato – the more productive the land, the higher the rent – got doctors to join the cause: potatoes were the healthiest food around and a cure for infertility, too. This was just propaganda. The potato benefited landlords because just an acre or two would suffice to feed a whole family. Population increase was good in the eyes of the Catholic church and so the clergy joined the potato craze.

For a while, this worked well. On the Atlantic coasts of Ireland and Scotland, there arose an industry in the early nineteenth century that offered a new source of income: harvesting and burning kelp to make fertiliser. This needed a workforce. Where grain had grown, landlords encouraged the crofters to grow potatoes because they demanded less space and much less labour than oats and barley:

the land could support more families and the field hours saved could be spent harvesting and processing kelp for cash instead. On the potato diet, populations rose to the highest numbers they had ever been. On the island of Tiree in 1830, there were more than twice as many people as there were at the end of the century. The new wealth, for a few decades, changed the face of Europe's poor western fringe. Then the potato blight hit.

A BLACK DEATH

People had known for decades that warm and damp weather in the early autumn was potentially catastrophic. It could suddenly turn potatoes in the fields into a liquid, fetid mush, with just a few blackened stalks pointing upwards. It's a smell that you catch tens of metres away and never forget – a stench that literally means mass death. In the mid 1840s that was not just of entire potato crops but of at least one million people in Ireland and a further 100,000 elsewhere in Northern Europe.

No one knew why or what did it, only that the black mould, which also affects tomatoes, could devastate whole fields in a few days. It spread scarily fast. Now we know the spores of the fungus, *Phytophthora infestans*, which is thought to have come from South America, spread on the wind. Blight outbreaks in 1845 and 1846 left the crofting families of West Scotland without a foodstuff that by 1840 was eighty per cent of their daily diet.

In Ireland, half the population was dependent on potatoes, most of them on just one particularly vulnerable variety, the Irish Lumper. Death and subsequent flight abroad in the 1840s and the years that followed, ultimately reduced Ireland's population by some three million. At five million today, it still has not recovered to pre-1845 levels.

For a long time, the only answer to the blight was growing more resistant varieties of potato, but such experiments could lead to more starvation if they were unsuccessful. In the 1880s, spraying crops with copper sulphate solution was found to be an answer. But copper was expensive. Despite the discovery, the west of Scotland potato crop failed again that decade, forcing more people to migrate across the Atlantic. In the First World War, copper was reserved for making ammunition casings. The resulting shortage of pesticide led to an outbreak of blight and a famine in 1916 and 1917 that killed 700,000 German civilians.

The disease still limits production in West Africa. Variants of *P. infestans* appear periodically, making 'late blight' an ongoing threat that could still lead to famine in the many parts of the world that still depend on potatoes. China and the United States have researched the blight's potential use in warfare: it is now officially classed as a biological weapon.

POTATOES IN THE TWENTY-FIRST CENTURY

China, a late adopter, is now the world's largest producer and potatoes are the world's fourth most popular food crop. Consumption, per capita, remains highest in some of the world's most vulnerable countries, from Rwanda and Belarus to Nepal, and farming everywhere is still highly dependent on artificial pesticides to tackle various blights, viruses and parasites that can destroy a crop. Even the threat of these can cause huge damage to the environment: for decades, US farmers sprayed cheap household paint on their potato fields late in the season to discourage insects.

Organic potato farming is difficult on a large scale. As with many vegetables, the chief input needed in the absence of artificial chemicals is for the grower to spend more time caring for the maturing crop. But potatoes face an unusual host of predators and diseases, which is why organically grown ones may be smaller and more blemished. They demand that customers are a bit more forgiving, ready to clean and cut out flaws, as our ancestors did. It is worth it: studies show that conventionally grown potatoes bring more fungicides and pesticides onto the plate than nearly any other fruit and vegetables.

Even peeled and washed, a United States Department of Agriculture study found eighty-one per cent of potatoes carrying chemical residues. So harmful can they be to human health, the UK government carries out quarterly tests on domestic and imported vegetables for pesticide residues. Its most recent report, from 2020, found the chemicals in 54 of 123 samples of British-bought potatoes, none of them at levels officially considered dangerous.

There are some hopes that preventative chemical dosing of potatoes against fungus and pests may end, as genetic research looks at the potato's natural defences against pests and predators. For example, like some other plants, including tomatoes and petunias, potatoes are now thought to be partly carnivorous. The hairs they grow on their leaves and stems trap insects, which die and decompose and feed nutrients to the plant. Harnessing and augmenting these abilities may help people use fewer pesticides.

Such adaptation of the potato might use the newer genetic modification (GM) techniques, known as gene editing, opening up

Reducing pesticide poisoning in Ecuador

In Ecuador's Carchi province, Farmer Field Schools have reduced terrifying rates of human pesticide poisoning. Continuous cropping of potatoes had produced not only high yields but highly favourable conditions for insects and fungal diseases, leading to massive applications of insecticides and fungicides. As a result of pesticide exposure, sixty per cent of people in the area showed reduced neuro-behavioural functions. Training enabled farmers to reduce agrochemical application costs – including fertiliser, pesticide and labour costs – by an average of seventy-five per cent with no effect on productivity. Follow-up studies showed that with reduced exposure to pesticides, local people recovered from the damage to their nervous systems.

Food and Agriculture Organization of the United Nations

the debate about whether these should be used on food for humans. Gene editing is still banned in the European Union for products for human consumption, under stringent existing EU controls on GM in food and crops. But Britain's exit from the union may permit a new approach, as was discussed in an August 2020 edition of *The Food Programme*. 'A blight-resistant potato would be such an advance in terms of reducing fungicide use,' said Guy Watson, the organic farmer who founded Riverford, the organic delivery business. He would probably accept it. 'I'm not blanket against [GM]. I'm not a Luddite. The organic community cannot be resistant to all technological advances, but I think we should be more wary of the unintended consequences.'

Micropropagation is a novel technique that reduces disease in potatoes without tampering with their genes or introducing chemicals. It involves producing plantlets and seed potatoes in sterile conditions, free of any earth that may contain disease spores or insect larvae. This means they can then be delivered to

Potatoes and the climate crisis

The greenhouse gas emissions of potatoes arise chiefly in the cooking of them. Unlike most vegetables we cook, they demand a lot of energy. Most damaging is oven cooking: a single oven-baked potato is responsible for 2,138g CO_2e (greenhouse gas emissions) – almost the same amount as it costs to bake two of them.

Microwaving the same potato will cost just 76g CO_2e. Cheese and butter – our favourite fillings – will add another 1,000g CO_2e. For comparison, a ham sandwich's greenhouse gas cost is about 678g CO_2e and it is suggested that we try to keep our daily emissions due to food to 3,000g CO_2e.* A filled baked potato for one will likely exceed that.

Oven chips are perhaps the worst form of potato dish: adding in processing, waste and oven use, they are responsible for greenhouse gas emissions of three times their own weight.

(* All such calculations are from S.J. Bridle, *Food and Climate Change without the Hot Air* (UIT Cambridge, 2020)

the farmer with a far better chance of growing without need for chemical intervention. It can also be done cheaply enough to help farmers in poorer countries.

There are programmes to help farmers trap and identify specific pests so that they can be targeted and controlled, rather than using blanket spraying of pesticides. Reviving the old field systems – used by both Donald Brown's family and the peoples of the Andes – is another popular idea. These used raised beds of manure and seaweed and other rotting foliage to plant potatoes. Modern farming ploughs up earth to do it, but that leads to soil erosion, rainwater runoff and the release of carbon into the atmosphere. Going back to 'lazy beds' is far more efficient, reduces blight and requires less land and artificial fertiliser. Sometimes the old ways do turn out to be the best.

POTATOES AND PLEASURE TODAY

From the food of necessity and hardship, the potato is now central to cooking for pleasure, not just as fuel, across the world. How many words are there for the magical transformations of a potato in the kitchen? A dozen or more come up without having to think: mash, hash, rosti, roast, soufflé, boil, croquette, crisp, dice, salad, chip. Not to mention all the potato flour breads, from scones to tortillas, and then *pommes dauphinoises*, where French potato cuisine reaches for the cream-laden divine.

The Irish, with the Spanish, were Europe's earliest and most enthusiastic adopters of the potato – it once defined their nation as beef did the British or leeks the Welsh. And it is a good bet that the Irish have more ways of enjoying potatoes than even the French, who were the first nation to explore the potato in cooking for the wealthy middle classes.

Long before the first French recipes were published in the 1800s, rural Irish people were eating more potatoes than anyone else. With the exception of rice in poorer parts of Asia, no one staple carbohydrate has been so central to a nation's diet. The chronicler Arthur Young toured Ireland between 1776 and 1778, writing about the economics of the poor – people who generally lived upon 'potatoes and sour milk … with, now and then, a herring'.

One family of six, he observed, could eat 252lb (114kg) of potatoes a week. It sounds like a vast amount. But it amounts to just a couple of medium-sized potatoes on each plate at a meal. It was not so poor a diet, either: 'The potato is the best single bundle of nutrition there is,' says potato historian John Reader. It could be boring, of course. The addition of those herring made a key difference, as Donald Brown found growing up in the Hebrides. In rural Sweden the same was true: a famous dish of sliced potatoes stewed with pickled young herring or anchovies is called Jansson's Temptation, possibly after a nineteenth-century pastor who easily resisted all pleasures of the flesh except this crusty, pungent dish.

The Irish have had much joy from potatoes. They are key to two nation-defining dishes: Irish stew and potato cakes (and their more pungently named siblings, boxty and fadge), as Darina Allen explains in her encyclopaedic *Irish Traditional Cooking*. Allen runs a restaurant, Ballymaloe House, outside Cork and her cookery school there is Ireland's most famous. In the book, she lists forty recipes in which the potato is the main item and many more in which it plays a large part.

Lots of the dishes are clearly born of the make-the-best-of-it pressures of subsistence living. There is 'Poor Man's Goose', suggested in a domestic economy schoolbook from the end of the nineteenth century. It is a sort of lasagne of pig lungs and other offal layered with sliced potato, dripping and a single sliced onion. That sounds tasty – it's a cousin to haggis – but other recipes smack more of desperation. Allen prints a recipe that 'can be used to make apple tart', bulking out the flour by adding mashed potato. Maris Piper is Britain's most popular potato today – Delia Smith's go-to variety for everything from mash to gnocchi. It is a mild-tasting variety, easy to grow and generally seen as a safe bet in the kitchen. But lots of people find Maris Piper, and others similar to it, boring and useless when in need of a potato with flavour for salads and roasting. For years, supermarkets resisted stocking anything beyond red potatoes and white: 'a potato is a potato' was the attitude Sheila Dillon found in 2009, reporting on big retail's resistance to oddly shaped or different coloured varieties.

That has shifted in the years since. Potato sales have been rising, boosted by growing interest in different flavours and the potato's amazing regional heritage. Of the many ancient and odd varieties that can now be found, Scotland's Kepplestone Kidney is one that chefs often pick when asked to choose a favourite. It has nutty tastes and an entrancing purple skin.

EPIC MASH

Mashed potato is a universal dish. The marriage of dairy or other animal fats – cream, cheese or butter – with broken-up boiled potatoes occurs across the world from the Americas to India. (In

China, Thailand and South East Asia, cooks tend to spice and stir-fry crumbly potato pieces to eat with rice.) Mash is comfort food for all of us. But it is not as easy as it sounds.

The tricky issues are around water content, the structure of the potato molecules and, of course, the ratio of fat to potato. French *pomme purée* takes all these to the max, with the ideal being a thick soup with huge butter content. The pioneering French chef Joël Robuchon is said to have got the dairy content of his silky-smooth puréed potatoes to an amazing sixty per cent by weight without the oil from the butter separating.

Crucially, Robuchon dries the cooked potatoes (the hard, waxy variety La Ratte, usually used for salads) over a low heat after a first run through a food mill or ricer. This gets rid of almost all the water – probably the most important detail in making truly legendary mashed potato. Then he whips in cold butter and boiling milk. But – as I know from experience – even following the YouTube video of him at work doesn't make this dish easy to get right. Get it wrong and it's an unusable, greasy mess.

Many people find these purées too gluey. Traditional British mashed potato is a fluffy thing with lots of air between light clumps of potato. Achieving that, while also adding copious butter, milk

'All the goodness is in the skin'

Not true. Potatoes contain carbohydrate, a little protein and nutrients including potassium, calcium and vitamin C: these are all mainly in the potato flesh. If you throw out the peel the only significant loss will be of half the potato's fibre.

and (as in my family) grated cheddar takes skill. One scientific tip, from the food writer Jeffrey Steingarten, is to let the peeled potatoes sit for a half hour or so in hot water (about 70°C) before they are boiled. This helps the starch in them gelatinise, producing a smoother purée. It also firms the starch molecules, making it harder to rupture them during mashing. Heston Blumenthal uses the same technique – and adds lime jelly at the end as a 'flavour bomb'. Indian mashed potatoes – *aloo bhartha* – made with ghee, mustard oil, green chilli and spring onion may be more exciting.

PERFECT CHIPS

French fries or potato chips – the name depends on where you're coming from – have elicited more gastro-scientific interest than any other form of cooked potato: their centrality in world-wide fast food ensures that. The key to the taste is contrast: that fluffy, light interior with a golden shell, crisp and sugary from the caramelising of the starches in the hot oil.

Most fried-potato lovers, from Thomas Jefferson (the US president who brought the idea home from Paris – hence 'French fries') onward, could argue for hours about what is the perfect way to fry a piece of potato. But it is more or less universally agreed that you need to fry them twice, except in Heston Blumenthal's case. He does it three times.

Gloriously different things all go under the name chips. There's many a shape and size, from bootlace and matchstick – dry and crunchy – to traditional British: yellow, shiny with oil and the thickness of a middle finger. The latter have never known a Maillard reaction (see chapter five): the rude might even call them soggy. In

Britain, we tend to cook our chips for longer, at lower temperature, than do the Belgians or Americans.

Beef dripping was once universal in chippies until a large part of the nation turned away from animal fats. South of a line drawn from Bristol to the Wash vegetable oil is usually used today – but north of that, the old fats still have their adherents. There is definitely more flavour in a beef-dripping chip and they usually win in blind taste tests.

One of the reasons fast-food outlet chips became so popular (and, for many, provide the ideal of what a chip should be) is because of the outlets' centralised preparation. Bags of part-cooked frozen chips are delivered to fast-food restaurants for another frying before serving: the system is convenient and economic but it is also key to a great chip.

162

The freezing process dries out the par-boiled chips and it turns out that frying them when frozen, rather than at room temperature, makes them golden on the outside (turning the starch into sugar) while leaving the inside pleasantly fluffy. You can do this at home – and most of us do. Forty years after the introduction of oven chips by McCain, seventy per cent of us now have a bag of them in our freezer.

The British, above all other nations, love potato crisps. Sixty per cent of children have them daily in their school lunch box and we collectively eat over six kilos of them each a year. They come from the United States – the first commercial crisp maker was George Crum, chef at a resort hotel in New York state's Saratoga Springs. One of its patrons, the shipping magnate Cornelius Vanderbilt, complained that his fried potatoes were too thick, so Crum tried slicing the potatoes as thin as he possibly could, frying them and salting them. Vanderbilt was delighted and 'Saratoga Crunch Chips' were born. Crum eventually opened his own restaurant and served the crisps fresh in baskets but he never packaged them or patented his idea.

At the end of the nineteenth century, crisps were sold in cardboard boxes but they still went off, or un-crisped, quickly. Airtight bags started to appear in the States in the 1920s. When advertising and marketing took on the snack – which provided great profits from a supremely cheap raw material – the word wranglers got busy. Professor Dan Jurafsky, a linguistics anthropologist, has studied the language of the modern American 'chip packet'.

'The longer the words on the packages, the more expensive the package,' he told Sheila Dillon in a 2017 edition of *The Food Programme*. 'The more expensive packages also used a lot of negation. That is to say, they told you what was not in the package: no grease, no fat. To be an expensive chip is to not be like those regular chips. With the inexpensive chips, there's more about authenticity, about being from an American region or the family farm, and "our founder, mom and pop". The expensive chips were authentic in a different way; they talked about the artisan and the work that went into how the chips were sliced, the selection of only the best potatoes.'

Henry Walker, a Leicestershire butcher, launched his crisps during the Second World War: now the company sells more than half the crisps eaten in Britain, making eleven million bags per day. That's not least because in 1954 Walker came up with cheese and onion flavouring. It is still Britain's most popular. But for all its Gary Lineker-fronted Britishness, Walkers has been owned by the global giant PepsiCo since 1989.

The crunching sound you hear in crisp adverts has a scientific purpose. By playing people different levels of sound as they ate identical crisps, researchers have shown that the louder the crunch, the fresher – and better – the consumer believes the crisp is.

Britain's crisp habit – the 184 packets we each eat a year – has its dark side. They are the biggest source of acrylamide (the particles that come from burning the sugars in foods) in the UK diet. Acrylamide levels in UK children are twice those of Denmark (they are also high in tobacco smokers) and the chemical is said – though this is disputed by the food industry – to be a significant cause of cancer. Some of the crisp-like snacks made from potato peelings and other highly processed vegetable starches have the highest acrylamide levels.

As you might expect, crisps are bad news when it comes to the environment, too. Every crisp is responsible for its own weight or more in greenhouse gas emissions – half of that because of the cooking, processing and packaging. Crisps that take seconds to consume come in a packet that may take a century to disintegrate. No bio-degradable version of the ubiquitous plastic packet has yet caught on – they exist but they are too expensive. In 2021, Walkers promised to change this by 2025.

THE HOMOGENEOUS STACKABLE HYPERBOLIC PARABOLOID

In 1956, Procter & Gamble, an American household goods firm most famous for soap, decided to get into the ever-growing savoury snacks market. They gave a brief to their lead research chemist, Fred Baur, to come up with a cheap snack that would address what market research said were the customers' key complaints about existing brands: crisps were not dependable. In one packet you could find broken, greasy or stale crisps among the good ones.

Baur came up with the hyperbolic paraboloid – the saddle shape – that had great internal strength but could be made very cheaply from dried starch paste collected from residues of processing corn, rice and potatoes. Not only would they be identical but they could also be stackable – and far less likely to break before the consumer got to them.

P&G were delighted with the idea. But they could not get the taste right. If they did, there were other problems: staff used as guinea pigs complained of diarrhoea attacks when different prototype flavours were tried on them. It took until 1968 before the Pringle – possibly named for Mark Pringle, the scientist who first worked out how to dry potatoes – was launched.

The first advertising slogan was 'Once you pop you just can't stop'. And it turned out to be pretty much true. Pringles were a huge success, soon marketed worldwide. So proud was Fred Baur, that when he died in 2008 his ashes were, at his request, buried in a Pringles tube ('Original Flavor').

But were they 'potato crisps', as it said on the tube? Other manufacturers objected. In 2008, a British court decided not, since the potato content was and is less than forty per cent and the saddle shape, said a judge, 'not known in nature'. But that was good news for Procter & Gamble because potato crisps attract value added tax and grain starch ones do not.

The Pringles' world-wide success has not won over all of us. Some flavoured varieties have more than a dozen different chemical ingredients and, like flavoured crisps, high levels of salt and sugar. (None as much sugar as the now discontinued 'peppermint white chocolate' and 'cinnamon sugar' Pringle flavours.) Since Kellogg's, now the owners of Pringles, have no factories for them in the UK, you can find tubes of Pringles that were manufactured in China or Indonesia: parabolic unbroken crisps come with a carbon cost, not just in the trans-world shipping but also the rigid tube (also invented by Fred Baur) and its plastic top.

Donald Brown in Tiree says he does not like Pringles, or potato crisps. But he still eats potatoes most days of the week. He buys his from the island supermarket, brought by ferry from the mainland. Only a few Tiree crofters plant them nowadays in the sandy soil: they are small, nutty and delicious, with a faint tang of the dried seaweed still used to fertilise the beds.

Sheep graze now where once the 'lazy beds' were and the machair grass grows richly on them. But when the sun is low you can still see the scars left by all that labour: parallel mounds laid out between the rocks wherever there was a morsel of land. At the height of the kelp and potato boom 200 years ago, 4,500 people lived on Tiree: today there are fewer than 600.

'How to make very good custards of potatoes' (1664)

Take a Quart of New Milk, six or seven Potatoes, boyled and very well broken, a couple of eggs beaten, sugar about a quarter of a pound, a little Nutmeg grated; mingle them well together, and put it into a shallow pewter or earthen dish, or else into Crust, having first put a little piece of Butter into the bottom, and so bake it in an Oven …; then keep it till it be almost cold, and you will have an excellent dainty and wholesome Dish, being both very pleasant to the Palate, and very restorative and strengthening to the body; and also so cheap that four pence charge, as much may be made, as will serve two reasonable men for a Meal.

A recipe from John Forster's *England's Happiness Increased*, a book addressed to King Charles II suggesting that potato growing was the answer to poverty in the land.

CHAPTER 7

A BIRD FOR EVERY POT

CHICKEN

'Most of us have forgotten the ways that chicken can actually taste … one of the costs of this mass market is a bird we have debased, we have to marinade, spice. We've accepted this in exchange for cheap chicken. When I went to Vietnam, I spent some time in villages where people are still eating local varieties of chicken that taste radically different.'

Andrew Lawler, author of *Why Did the Chicken Cross the World? The Food Programme*, 2015 f you were to design a creature for human use, you might come up with the chicken: easy to keep, self-feeding, useful when alive (as a guard and yard cleaner), tasty and nutritious when dead. No meat is quite so astonishingly quick and cheap to turn from newborn into a family's meal. A better friend to us, you might reckon, than the dog.

For ten millennia, the chicken has lived alongside humans: turning our leavings into protein, acting as a burglar alarm, a clock, predictor of fortunes, provider of useful materials – from feathers for insulation to bones for tools. Also, of course, it gives food, alive and dead. Each female comes ready to produce several hundred eggs: long-lasting packages of nourishment that can be turned into a breakfast staple, a whole world of cakes, meringues, soufflés and sauces. And at the end of that work, you eat the machine that gave you these things.

We have repaid all this by making the chicken a success, in the most basic way a species can be judged. It has partnered with us so brilliantly that it is now the most populous bird there is. At any moment, there are some thirty billion chickens alive, far outstripping the next most populous species, the red-billed weaver bird of sub-Saharan Africa, which numbers some 1.5 billion.

But the weaver will live several years; almost all chickens reared for meat live less than two months. Most male chicks are killed on hatching. If you were one of those birds, whose normal life could be a decade or more, this fate might make your population explosion seem less of a triumph.

When we say we love a food, or indeed a domesticated food animal, we're indulging a hypocrisy that many people find increasingly hard to stomach. The British love of chicken has seen it turn from a treat – something that in 1950 we ate less than once a month – to being an everyday staple, the most-eaten protein of all, twice as popular as beef or pork.

It's not hard to put together a case that, after all those years of mutually beneficial coexistence, humans have cruelly betrayed the chicken. We have transmuted it into something nightmarish: an organism treated more like a plant than a sentient being.

CHICKENS AND US

Humans first caught and bred chickens not for food but for entertainment. One male jungle fowl cock will fight another to exhaustion or death for supremacy, the right to rule the roost. Thus, cockfighting, and gambling on it, became the first chapter of the story of our attraction to chickens. The dead birds of the fights were eaten, of course, while the winner went on to breed, passing his superior fighting traits to another generation.

So chickens became players in the everyday drama of human life, living alongside us, pecking at our scraps for nourishment, prized for many things. Part of the attraction is that they are low maintenance: unlike mammals, the chickens' young can look after themselves almost from hatching.

They are clever, too: they recognise their carers and communicate among themselves with as many as thirty distinctly different clucks. Roosters or cockerels – the male birds – have been both alarm clocks and alarm systems, the first to warn of intruders. On scraps and farmyard pickings, female birds will produce packaged protein – perhaps 600 eggs each, though the longest-lived may be capable of well over 1,000. From their feathers to their bones, chickens were useful to our ancestors: across East Asia their feet are a delicacy too. Chinese chefs fry them for several hours to make *fèng zhǎo* – 'phoenix talons'.

172   And so chickens clucked their way into our hearts. They have rarely been loved as dogs or cats are, but anyone who has lived with a chicken in the yard has a respect and fondness for them, their habits, their curiosity, their complex family life. Chickens have entered our rituals and our metaphors. A mother hen is the epitome of fussing and caring; a cockerel – France's national emblem – a symbol of vainglory, foolish bravery, masculine dominance. The bird appears on the walls of Egyptian pyramids and on English church steeples. Chicken guts, cast upon the ground by priests, could deliver messages to politicians and generals of Ancient Rome to guide the fate of the empire.

But chickens' role in human society has recently and radically changed, as scientist Gregor Larsen told the BBC. 'The last fifty years is a complete departure from our association with them over the thousands of years before … When people think of chickens now, it's just a homogenised bird with no variation – we don't see them, just bits and pieces in a supermarket. But people who have backyard chickens see them as having personalities, as sentient beings.'

Since chickens became docile, protein-fruiting plants, the metaphorical use of them has changed completely: now we say chickenshit, bird-brained – a coward is a chicken. We lost our respect for them.

THEY DON'T TASTE LIKE THEY DID

The British and the Americans like to sit down round a whole chicken: more than a treat, it is a symbol of family togetherness. In the 1920s, this notion became a political symbol: 'A Chicken for Every Pot', promised campaign ads for the (successful) Republican presidential candidate, Herbert Hoover. This much-mocked promise was in fact echoing a medieval French king's desire: 'I want there to be no peasant in my realm so poor that he will not have a chicken in his pot every Sunday,' said the reformist Henri IV.

Before industrialisation, most ordinary people never ate a chicken other than an exhausted laying hen or a cock who'd lost his position at the top of the flock. Older chickens have their advantages, taste-wise. Classic recipes for Jewish chicken soup call for 'an old scrawny haggard chicken'. This would be boiled for twelve hours to make a gorgeous, golden, unctuous stock ready for matzah balls or noodles. The meat is discarded – they say that if you've boiled the chicken properly even the cat won't want the remains. But nowadays it's hard to find an old chicken, kosher or not. When I asked at Bower's, the specialist butcher in Edinburgh, they could think of no way of getting hold of one other than going to a traditional, truly free-range egg farm. There are not many of those.

The British and the French still like to roast or stew chicken whole but in most of the world the carcass is chopped up and sauced, spiced, stewed in wine and herbs, or fried with seasoning and flours. Recipes today are all about putting flavour back into something that, because of its short life, has very little of it – a chicken sandwich doesn't work without mayonnaise, butter or at the very least some salt.

Even before chicken was industrialised, it was not prized for its taste. 'Poultry is to the chef as a canvas is to painters,' wrote the first foodie, the eighteenth-century gourmand Jean Anthelme de Brillat-Savarin. In his era, goose, duck and wild birds were seen as far superior. Classic French recipes using a whole chicken always come with a command to baste them in goose or even bacon fat. In sixteenth-century Italy, chicken was stuffed with nuts and served with sauces of lovage, pepper, oregano and honey.

174

People had known how to breed them for meat as well as fighting for millennia: in Ancient Greece the practice of caponising – castrating the male bird so it grew bigger – was common. 'Spring chickens', prized for delicate flesh, were capons: nowadays male birds produced by the egg-laying breeds are discarded after hatching, not worth feeding until big enough to eat.

BAKING NOT ROASTING

Roasting the whole bird is tricky: exposing a delicate meat to high temperature is a gamble at the best of times. A whole carcass contains thin and thick muscles, the very different meats of leg and breast, skin: all parts that need specialist attention.

The food laboratory run by Nathan Myhrvold (which also baked 36,000 loaves in a quest to understand bread, see chapter

A perfect roast chicken?

This is one of cooking's most lively debates. Every traditional chef with a book on the market has a tip – some of them wildly contradictory. Simon Hopkinson starts the bird at 230°C, Pierre Koffman at 180°C. But there are some well-worn ideas:

- 'It's not the heat, it's the humidity.' Use the cavity. Thyme, onions and half-lemons all serve to retain moisture and add flavour.
- Steam, from water and wine, reduces cooking time, keeps things moist and helps brown the skin.
- Slow-roasting at 100°C can bring out remarkable flavour. Use a probe thermometer to cut out the guesswork. If the bird hits 65°C in the centre of its breast for 12 minutes, it's safe.
- Buy a quality, long-lived bird; the flavour improves the longer the bird lives. 'Organic' gives some guarantees. The stock from the bones will be better, too.

Variations on these themes include:

- Making a tarragon and butter mash and slipping it between the skin of the breast and the flesh.
- Giving the bird a start by poaching it briefly in white wine.
- Taking the temperature right down and slow roasting it in a heap of moisture-laden onions and root vegetables.

one) has looked in obsessive detail at what happens when we roast a bird: essentially how we manipulate heat, convection and humidity. So little does the ordinary cook understand about this, Myhrvold points out in his book *Molecular Gastronomy*, that we are not even aware that the bird does not roast in a kitchen oven but bake. Roasting is holding something to radiant heat, like a flame or an electric element. Baking uses ambient heat to allow the water at the bird's surface to boil and evaporate.

Cooking a chicken well in this way is about the humidity in the oven, not the heat. 'Water vapour is the primary cause of erratic results,' the Myhrvold report says. Because water conducts heat much better than air, the amount of liquid will determine how your bird cooks. Resting the roast chicken is a myth, he goes on, but basting it with oils from the pan definitely helps because the evaporating super-heated oils cause explosions, which in turn increase water vapour circulation, delivering a golden-brown crispy skin.

'MEAT FOR THE PRICE OF BREAD'

The industrialisation of chicken began in the mid-twentieth century. Selective breeding produced chickens that would fatten at extraordinary rates or produce far more eggs, but it came at an enormous price – to the chickens' welfare and to the environment. No more farmyards: confined to barns, even when 'free range', the chickens may be too unaccustomed to daylight to venture outside.

The change in our relationship has been mathematically enormous: in 1950, the UK slaughtered one million chickens and in that whole year the average person ate a kilo of chicken meat. By 2015 we were slaughtering one billion annually and eating half a kilo of chicken each week. Chickens themselves had changed through breeding: top heavy (because of the value of the breast), fast growing and incapable of the social and foraging behaviour that they will show if released into a normal, non-intensive environment.

Today, it takes just six weeks to turn a 40g newborn chick into a 2kg broiler chicken, using rather less than twice that weight difference in feed. (A beef cow needs six or seven times its weight in feed.) This process began in the 1940s, when the American poultry industry launched a marketing campaign under the slogan 'Meat for the price of bread', with a high-profile competition for farmers to breed a 'Chicken of Tomorrow'. One of the marketing department's key concerns was that a whole bird of the time was too much for a couple but not enough for a family Sunday dinner. What was wanted was a 'sumptuous chicken', a bird with a breast like a turkey's, as Maryn McKenna explains in her history of the US industry, *Plucked!*

They got it, and at a good price too: it's said that in 1920 an American worker on national average pay had to put in two hours to earn enough to buy a chicken; today, it is just eighteen minutes. In 1967, a kilo of chicken meat cost 39p in Britain, according to government records: that is £7.24 today when adjusted for inflation. But in 2021, a kilo of meat cost £2.76 and a whole chicken, at Asda and Tesco in November 2021, just £2.66.

More of the chicken is now put to use. In 1963, a professor of agriculture, Robert Baker, came up with a way of turning the waste from the factory – necks, spines, skin, mechanically recovered meat scraps – into fryable cakes that Baker first called 'chicken franks' and 'bird dogs'. Now we know them as chicken nuggets: they are the cheapest meat protein meal of all.

But the super-chicken that selective breeding produced is a very long way from its wild ancestor, the jungle fowl, and quite different too from the busy, free-roaming hen of children's picture books or *Chicken Run*. This one looks more like 'an olive on two toothpicks', in McKenna's phrase: it can no longer perch above the ground and may be so top heavy it can only sit on the ground, in its own waste.

Since most consumers prefer breast meat, that part of a broiler is now twice as big as on a traditional breed, as much as twenty per cent of its total weight. The legs may be malformed and unable to support the bird for long. In the last hundred years, the average chicken slaughtered and ready for sale has gone from an average 2.5lb (1.13kg) to 6lb (2.7kg). But they have got cheaper to produce: it used to take 4lb of feed to produce one pound of flesh: now it takes less than two.

In the most productive American battery farms, chickens – whether for eggs or meat – live in an area the size of an A4 piece of paper, in cages stacked nine or ten high. In Europe, things are better: the EU banned the most restrictive battery cages in 2012 and there are moves to get rid of all cages in chicken farming by 2027.

Boiling an egg

Temperature is the key. They don't have to actually boil: a hard one only needs to get to 80°C; a soft one to 65°C. But it'll take half an hour to do it. Try putting an egg in cold water and bringing it to the boil – and then removing the pan from the heat. Cover it, and six minutes later you *should* have a perfect soft-boiled egg.

To live so unnaturally is damaging to more than just the chicken. Farming poultry cheaply enough to satisfy modern consumers' demands has damaged the environment and set a bacterial time bomb that could drastically alter the future of human health (an issue discussed later in this chapter). More antibiotics have been fed to chickens than all other animals, including humans, put together. The cost in fossil fuels to produce fertiliser, wheat and soy for feed makes the world's cheapest, most popular meat very expensive indeed in the alternative accounts book that tallies the true cost of food to future generations.

DIRTY CHICKEN

In 2014, the *Guardian* newspaper published the gruesome results of an undercover investigation at two processing plants of one of Britain's biggest chicken producers, 2 Sisters Food Group. Snatched video footage and the testimony of former workers told a horror story: 'pile-ups' of offal on the factory floor, traces of faecal matter on birds ready for packaging, the washing tanks for newly killed birds not emptied or cleaned for days, plucked carcasses falling into the dirt and simply being picked up and hung back on the processing line. In reply to the *Guardian* piece, 2 Sisters issued this statement: 'The allegations about our processing sites at Scunthorpe and Llangefni [Anglesey] are untrue, misleading and inaccurate. Both sites have British Retail Consortium 'A' grade Food Standards certifications, based on a number of announced and unannounced visits. In addition, we and our customers carry out a number of audits of our operations. None of these audits have uncovered any concerns about our hygiene standards or food safety'. However, the newspaper connected all this with government statistics showing that two thirds of chicken sold in Britain was contaminated by the time it reached the shops, mainly with the dangerous bacteria campylobacter.

From leading supermarkets to fast food chains, British chicken eaters can hardly avoid the 2 Sisters Group. From small beginnings in meat packaging in the Midlands, the company started acquiring others in the early 2000s. By 2013, it was processing nearly half the chicken in the UK. It now has plants in Poland and the Netherlands and supplies one third of all chicken eaten in the UK.

The Boparan family, who still ran the company in 2022, could not be more honest about what they do: 'We buy for less, produce for less, sell more for less.' 2 Sisters' strategy? To become the world's leading food company, by doing things as cheaply as possible. Today, they produce ten million chickens each week.

Not many journalists get into industrial food factories, let alone ones where animals die. The practices on view are simply too disturbing for the public to know about. But after the alleged hygiene failings and resulting investigations of 2014, *The Food Programme*'s Dan Saladino managed to accompany a Food Standards Agency inspection of the 2 Sisters chicken processing plant at Scunthorpe. It is the UK's biggest – processing 1.75m birds a week.

Saladino watched as lorries arrived, each carrying 6,000 live birds. The system seemed highly efficient: the live birds are offloaded into a 'Wendy house' where fans keep them cool and then they are stunned with gas. It all goes at remarkable speed: 180 birds go through a minute. Unconscious, they are hung upside down on shackles on the beginning of an overhead monorail, which transports them through the factory and the entire process.

First stop is the automatic kill blade, which cuts their throats so they can 'bleed out' into a tank. Then they go into a pool of hot water. This opens their pores so a machine can beat their feathers from them. A river of blood, water and feathers streams below. The feathers go to landfill, incineration or into animal feed. They may even be fed back to chickens.

It all sounds very efficient, but when the defeathering machine beats the un-gutted chickens it also spreads germs – faeces are squeezed from the carcass and salmonella and campylobacter bacteria may spread to the empty feather pores. Although the bug is killed by thorough cooking, one of the reasons customers are advised not to wash chickens before cooking them is for fear that the bacteria on the skin will travel to the meat.

The chickens' journey through the plant continues: meat health inspectors and vets inspect them, but of course bacteria cannot be seen with the naked eye. Next, the birds queue for the 'evisceration department', where guts are removed, heads and feet taken off. Then the birds are unshackled, checked for stray feathers, faecal matter and bruises, and then chilled and packaged for shipping to Britain's supermarkets and restaurants.

Saladino took some of the worries about this process, raised by a Parliamentary committee, to the company's top management. They were defensive. Technical director Simon Hewitt rejected the suggestion that the 'line speed' – the rate at which the birds went through the process – needed to be slowed down so inspectors had more time and fewer mistakes occurred. 'Our equipment runs at the line speed specified by the manufacturers,' he said.

Hewitt promised new investments in hygiene, including a novel process involving chilling the fresh chicken to minus 2°C.

A pilot study done at the Scunthorpe plant showed this would reduce incidence of 'highly contaminated' birds leaving the site to a mere 2.2 per cent.

The bottom line, it was clear from Saladino's interviews, was that if the British public wanted cheap chicken they would have to put up with more than one in fifty of them coming to the shops carrying significant amounts of food poisoning-causing bacteria. In fact the risk to consumers may be higher. Today the National Health Service website advises treating raw chicken with extreme care, stating that, according to surveys, more than fifty per cent of birds sold are contaminated with campylobacteria.

After Dan Saladino's visit to the Scunthorpe plant, it was less than a year before 2 Sisters was in the headlines again. In 2017, another undercover investigation at the West Bromwich plant filmed more breaches of hygiene rules and, more importantly, found staff changing labels recording 'kill date', allowing the 'use by' date seen by consumers on the packaging to be put back. In a statement to ITV and the *Guardian*, 2 Sisters said it took the allegations very seriously: 'Hygiene and food safety will always be the number one priority within the business, and they remain at its very core ... we are never complacent and remain committed to continually improving our processes and procedures. If, on presentation of further evidence, it comes to light any verifiable transgressions have been made at any of our sites, we will leave no stone unturned in investigating and remedying the situation immediately.'*

---

* You can view the full statement at https://www.theguardian.com/business/ 2017/sep/28/uks-top-supplier-of-supermarket-chicken-fiddles-food-safety-dates

CHEAPER AND CHEAPER

More and more of these problems emerge in the cheap meat factories. There is a scandal around industrially produced protein every year – horse meat in burgers, salmonella outbreaks, dioxins in pork and farmed fish. The root of problems, analysts and farmers agree, lies in meat being too cheap. 'We're working with a food sector that has a two to three per cent profit margin,' Professor Chris Elliott of Queen's University Belfast told Saladino. 'Can you imagine if you try to operate a business on a two per cent margin? If you have a bad day that can wipe out your profit for a whole month. Two bad days and you're out of business.'

In the UK in 2020, according to the World Economic Forum, we spent 8.2 per cent of household income on food, though since our exit from the European Union food prices generally have been rising. Twenty years ago, it was fifteen per cent; forty years ago, twenty per cent. So in real terms we spend much less on food than we ever have in the past. In Europe, generally the number today is twelve to fifteen per cent. 'Britain has some of the cheapest food in the world, beaten only by the United States – where just 6.4 per cent of a household's money goes to the food bill.

That is not good for health or animal welfare. 'The pressure goes onto the producers. It's always about cutting costs,' says Elliott. It is hardly surprising that outbreaks of bacterial poisoning continue, despite the outrage in the media and Parliament that met the scandals of 2012 and 2017.

A worrying new issue is that recent cuts in the budgets of the Food Standards Agency and in local public health policing mean that problems may only be noticed and analysed when significant

Chickens, eggs and the climate crisis

Chicken is far less costly in terms of greenhouse gases than other meats and dairy, largely because of its very short life. It is half the cost of equivalent amounts of pork or cheese, and twenty per cent that of beef. But it is still expensive in terms of damage to the planet. Buying more smartly can reduce that.

The principal cost of eggs and chicken meat in climate-altering terms is through deforestation to produce soya for feed. For a hen to lay twelve eggs she needs to consume roughly 1.8kg of feed. Most of the three million tonnes of soy we import to the UK annually goes to chicken feed, and less than thirty per cent of it is certified as sustainable. So consuming chickens and eggs produced with locally, sustainably sourced feeds can make a big difference.

The overall greenhouse gas (GHG) emission of an egg is five times its own weight – and about the same as that of 30g butter. The GHG cost of boiling an egg is a third of that of producing it. But the GHG cost of scrambling two eggs with butter comes to 1,174g CO_2e – more than a third of the recommended daily expenditure of GHGs on food per person and three times the emissions of a single boiled egg.

Figures from S.J. Bridle, *Food and Climate Change without the Hot Air* (UIT Cambridge, 2020.

numbers of people are affected. 2020's salmonella outbreaks killed at least one person and put several hundred more, half of them children, in hospital.

The source was found to be budget chicken nuggets and other ready-to-cook products sold at Lidl, Aldi and Iceland. The source of the contamination was traced back to a factory in Poland. Early in 2021, a brand called Southern Fried Chicken's take-home boneless chicken pieces were recalled by Sainsbury's and other shops after another salmonella scare.

The big processors in the UK, like 2 Sisters, Avara and Moy Park, argue that, despite the headlines, things have improved remarkably in recent years – for humans who eat chickens, if not the birds themselves. After all, of the more than three billion chickens slaughtered in the UK between 2016 and 2018, 'only sixty-three million were rejected as unfit for eating at the processing works – just over two per cent,' according to the British Poultry Council.

That is a sign that more rigorous inspection regimes are functioning, according to the industry. In the average battery farm, four per cent of the birds are killed because of disease or injury. Of course, critics see these statistics as a scandal, not an achievement. 'These figures shine a light on the very poor conditions that are the norm in Britain's poultry sector. Most broilers, turkeys and ducks are farmed in crowded, stressful conditions that make them vulnerable to disease,' says Compassion in World Farming's Peter Stevenson.

ANTIBIOTIC OVERLOAD

There is one crucial improvement that the industrial processors of cheap chicken have initiated that will certainly have a far greater impact on human health than any project on faeces-contaminated meat. That is the reduction – at long last – in the use of antibiotics in mass poultry farming. This practice, says Maryn McKenna, has been 'the greatest slow-brewing health crisis of our time'. In 2016, the United Nations General Assembly declared antimicrobial resistance the most urgent risk to human health, development and security: 'a fundamental threat'.

Antibiotics were first added to animal feed in the United States in the late 1940s to counter the problems that emerged from packing chickens unnaturally close together in the new battery systems. In the same era, scientists were confirming that overuse of the anti-bacterial wonder drugs that had transformed human health – to the point where an infected wound no longer meant a significant risk of death – might be a problem. Quite simply, antibiotics lost their efficiency because the bacteria they targeted evolved to develop defences to them: this was antimicrobial resistance (AMR).

Yet, just as alarm bells were being rung over AMR, the feed industry started profiting from a surprising side effect of the drugs. The antibiotics made the chickens grow faster – three times as fast – for reasons nobody quite understood. So, from the 1950s, antibiotics started to be added to feed for chickens and pigs as 'growth promoters'. At the height of the antibiotic craze in the late twentieth century, sixty per cent of the world's production of the drugs was being fed to chickens.

But of course, the bacteria in the animals' stomachs started adapting to cope with the onslaught of the drugs. This inevitably meant new, drug-resistant bacteria, including salmonella, were being consumed by humans, who in turn lacked resistance to the

Fake eggs

This book is full of stories of attempts to make more money out of food by hoodwinking the consumer but the story of the Chinese fake chicken eggs is perhaps the strangest. Some time in the 1990s, fraudsters in China developed a technique to mould the yolk, white and even the membranes of an egg out of gelatine, benzoic acid, food colouring, calcium chloride and other substances. Paraffin wax made a convincing white shell. The fakes not only weighed and felt the same but when cracked into a pan, fried like chicken eggs, though a bit 'bubbly'. Once trained, a person could make 1,500 a day – and make a profit selling them at half the price of real ones. The only problem with the eggs was that when eaten they caused stomach aches and memory loss. Nevertheless, fake eggs spread across South East Asia.

new strains. By the mid-1950s, doctors were already finding that antibiotics they had used successfully against common infections were no longer working.

Today, AMR is blamed for 700,000 deaths worldwide a year, costing billions in healthcare costs, lost wages and productivity. Unless solutions are found – and there's no lack of scientists looking for them – by 2050, ten million people will be dying every year as a result of antibiotic resistance, according to a major review published in 2016 in *The Lancet*: three times as many as died in the initial eighteen months of the Covid-19 virus. The economic cost is put at £100 trillion annually.

RESISTING THE SCIENCE

By the 1970s, as McKenna puts it, 'chicken had become dangerous' in its mass-produced form and was recognised as such by scientists across the world. Big agriculture and the pharmaceutical industry successfully fought off attempts to regulate the use of antibiotics, even as growth promoters. In the United States in the 1980s, corporations continued to produce and use new antibiotics in animals without even testing the risk posed to humans. This meant that the new, more powerful antibiotics developed specifically for humans to counteract drug-resistant bacilli were used in animal feed in the same way that had made the previous drugs useless.

The research seemed unarguable and the evidence kept mounting – not least because farmers themselves found antibiotic after antibiotic becoming ineffective. Some European countries began limiting antibiotics in feed in the mid-eighties. The European Union brought in its first rules in 1999, banning antibiotic use other than for disease control in 2005.

But it was not until then, after a round of lawsuits, that regulators in the US managed to force the first removal of an animal drug from the market because it would cause resistance to antibiotics and so harm humans. From 2022, no antibiotics at all can be used in animal feed in the EU. But campaigners fear that Britain will be more exposed to antibiotic-laced meat as, post-Brexit, imports increase from the United States.

Use of antibiotics as preventatives and as growth promoters in animals will continue in many countries, but in most of Europe and in North America the actual quantities used have declined by as much as eighty per cent or more – or so the industry bodies claim.

Many big producers, including 2 Sisters, say they have committed to plans to continue to reduce their use of antibiotics.

Perdue, one of the United States' biggest producers, now cites 'No Antibiotics Ever' on all its chickens – 'no exceptions and no fine print'. But there is an 'animal welfare' get-out. If a flock becomes diseased, antibiotics can be used – even organic chicken permits that. During the coronavirus lockdowns of 2020 and 2021, chicken sales collapsed as restaurants closed and home cooking habits changed. The big farms, whose output is carefully calibrated to usual seasonal demand, began to experience huge overcrowding problems. Fearing disease outbreaks and bankruptcies, British authorities suspended some rules on antibiotic use for farmed animals from chicken to salmon. Even in normal times, over sixty million chickens a year are still rejected at slaughterhouses in England and Wales because of disease and 'defects'.

Reducing antibiotic use

British Poultry Council Chief Executive, Richard Griffiths, says: 'The British poultry meat sector's drive for excellence in bird health and welfare has been delivering responsible use of antibiotics and safeguarding the efficacy of antibiotics across the supply chain. We've successfully reduced our antibiotic use by eight-two per cent in the last six years and have stopped all preventative treatments as well as the use of colistin [a crucial 'last-ditch' drug in treating human infections]. The highest priority antibiotics that are critically important for humans are used only as a "last resort".'

A PERFECT BIRD FOR SUPPER

The bird is crispy brown on top when we raise the pan lid. It smells of thyme, garlic and that nose-lifting, unbeatable chickeny scent. The breast meat is creamy-white but textured – it has a chew to it. The dark meat of the legs is almost purple and it has a nutty, rich taste: these are muscles that have worked. This chicken has walked, foraged, fought, roosted: it has lived.

It is indeed special. It is one of Linda Dick's chickens, famous throughout Scotland. The bird cost about four times as much as a standard supermarket broiler and it lived nearly three times as long. It deserves a debate over how to cook it. In our family, this happy argument comes down to two well-loved ideas: is the bird going to be pot-roasted in milk, spiced with garlic, nutmeg and lemon (a treasured recipe from the food writer Diana Henry)? Or steam-roasted in the oven?

For thirty years, Linda has spent every weekend plucking chickens. That is not something human beings usually do nowadays – there are machines to take the feathers off them, machines to clean and gut them, and package them too.

But Linda, who has a small mixed farm in south-west Scotland, likes to produce chickens that taste good and can provide a family with two or even three meals: chickens like the ones she knew as a child. So the Dick chickens live and die as chickens used to half a century ago, before they became the cheapest meat on the planet. 'It's simple. We do it exactly like my uncles did when I was wee,' she says.

These chickens live what you might call a good life, unlike almost all the ninety billion of their species that die every year for human consumption. Linda buys chicks from one of Britain's

How do you know an egg is fresh?

The question has been troubling mankind since we first wondered which, chicken or egg, came first. It's important not just from the point of view of health but also because eggs will do different things at different ages. Newer eggs will fry better, with more shape, and their yolks make a stiffer mayonnaise; some baking requires older eggs.

Since the introduction of the British Lion stamp after scandals around salmonella poisoning, it has been easy to tell an egg's age: it was laid thirty days before the 'use by' date (though, if refrigerated and uncracked, you can eat it later). Fresh eggs will sink in a bowl of water, less fresh will float because the air sac inside them gets fuller with age. But floating eggs aren't necessarily inedible. The oldest testing tool in existence is your nose.

largest suppliers. Chickens in industrial farms are no different, genetically, from hers. She simply rears them as it used to be done. The key difference in taste, she thinks, is the slow-rearing and the fact that the birds are hung in the traditional manner for a few days before cleaning is finished and they are sold. The hand-plucking makes this safe.

Linda Dick's meat chickens live for four months or so, roaming in and out of their barn. In his book *In Search of Perfection*, the chef Heston Blumenthal obsessively researched the best chicken for roasting. Linda Dick's made his list – the others were Ellel Free Range Poultry from Lancashire, Label Anglais (an attempt to replicate the French Label Rouge in Essex, now renamed English Label) and Waitrose's own-brand organic. He, like so many other converts, wonders why more producers are not attempting premium, healthy, well-treated birds. 'It's how chicken is meant to taste and, I reckon, a far better eating experience.'

Linda mixes her own feed, from vegetable scraps and grains. It is not designed to make them fatten as fast as possible but at something more like a normal rate. Her flock very rarely needs intervention by the vet, and never any antibiotics, because they do not live close packed in a way that encourages disease. At the end of their lives, they sell in butchers in Edinburgh, the Lothians and Borders for three or four times the price of an average supermarket chicken, yet Linda and her husband can never produce enough of them. 'I don't really want publicity,' she says when journalists ring up. 'I can't do any more than I'm doing.'

CRY FOR THE CHICKEN

The chicken's twentieth-century fate used to inspire horror stories. In her dystopian-future novel *Oryx and Crake*, Margaret Atwood pictured the genetically modified ChickieNob, an industrial chicken that was just mouth, digestive tract and twelve breasts (or drumsticks, depending on your fancy). But, in recent years, consumers seem to have turned against ever-cheaper factory chicken on health and welfare grounds.

The promise on a label of antibiotic-free chicken inside does seem to have marketing potential – the most likely means to seeing a change in the industry's practices. That worked with the terms 'free range' or 'grain fed' – though in practice, the latter mean little. (All chicken is grain fed nowadays and free range may mean simply a door being open at the end of an enormous barn.) There is scepticism, in retail, with good reason. Despite years of negative publicity about cheap, industrialised chicken, only around two per cent of British chicken sold is organic.

The key is always price. It is very hard to persuade British consumers, or American ones, to pay the extra that could permit a chicken a better life, or even better feed – to convince them that a chicken should cost more than bread. In January 2008, Channel 4 spent a fortnight airing programmes by campaigning chefs Jamie Oliver and Hugh Fearnley-Whittingstall devoted to the problems – taste, health and welfare – of a national addiction to cheap chicken.

In *Hugh's Chicken Run*, Fearnley-Whittingstall was reduced to tears as he looked at how cheap chickens are kept. *Jamie's Fowl Dinners* showed a dinner party of celebrity guests exactly how chickens are stunned and killed. He introduced viewers to MRM (mechanically reclaimed meat), a grey-pink sludge made from chicken carcasses after butchering, which goes into cheap sausages.

Over three million watched the shows and they reacted: organic and free-range sales soared thirty-five per cent, while cheap chicken dropped by seven per cent. The RSPCA reported that in a poll of British people, seventy-three per cent had pledged to buy better-produced chicken – and worried there was not enough to supply that demand.

But six months later, everyone's outrage had diminished and statistics on chicken sales showed that the market had not shifted at all. Since then, change has not been impressive. You can still buy a whole chicken for £3, less than a McDonald's Big Mac. In

2020, organic birds were just 2.1 per cent of the one billion British meat chickens eaten, up from one per cent eight years earlier. It is a sad truth that, despite the historic cheapness of British food, most consumers – according to surveys – will still think of price before taste and welfare.

In the hard times of 2022, with food price inflation hitting levels not seen in twenty years, even the kings of cheap chicken, 2 Sisters Group, believe things have to change. 'How can it be right that a whole chicken costs less than a pint of beer? You're looking at a different world where the shopper pays more,' said Ranjit Singh Boparan in October 2021.

As owner of both Bernard Matthews and the 2 Sisters Group, his factories are now the source of one third of all Britain's poultry. Boparan said he expected the price of chicken to rise ten per cent over the winter of 2021/2022. Food in Britain, he added, was 'too cheap': the system needs a 'reset'.

Linda Dick thinks her business will come to an end when she retires. She's never succeeded in finding anyone prepared to take on the labour of hand-rearing and hand-plucking, even though her chickens sell for three to four times the price of an ordinary supermarket one.

This is all rather different in France, where flavoursome, better-treated chicken is valued. Shoppers are used to spending as much as €50 for the appellation controlée *poulet de Bresse*, a delicacy beloved of Michelin chefs and famous in the country for 500 years. These live outdoors for a minimum four months, fed on corn and milk products, with a guaranteed space per bird about two-thirds the size of a car parking space. A bit cheaper are the government-regulated Label Rouge 'pasture-raised' heritage-breed chickens.

These are slow grown – they may live for one hundred days, rather than the thirty-eight of a cheap meat chicken – and with other standards close to organic ones in the UK, they cost twice as much as intensively reared French chickens. But Label Rouge has nearly two-thirds of the French chicken market.

TOMORROW'S BIRD

Chicken is becoming both more *and* less planet friendly. It is still growing in popularity world-wide, particularly in developing countries, where people invariably increase the amount of meat in their diets as they become richer. That, in turn, increases climate-damaging emissions. But switching from pork or beef to chicken causes fewer emissions, not least because chicken needs less feed per kilo of meat.

In the world's richest countries, where we still eat two or three times as much meat as in less developed ones, some things are improving. The practice of feeding climate-damaging foods to chickens, like soya, is under question, just as was the cheap fishmeal diets of earlier generations of factory chickens. Chickens need protein and fats but there are other ideas for sourcing these, not least the use of insect larvae, which can be produced with a very low greenhouse gas cost.

Anyone disgusted by that should remember that chickens have been picking around for food in farmyards and compost heaps for centuries – they will catch and eat mice if they can. It is clearly better for a chicken to eat locally produced maggots than soya from Brazil. This is not just about the good of the planet: corn- and soya-fed chickens are actually less healthy for us. It's one of the reasons omnivorous humans are in need of omega-3 oils: our favourite bird lacks them because its diet is now almost entirely vegetarian.

For consumers who care about animals' lives, spending a little more is the most obvious answer. You get a bonus in better taste and in 'nutrition density', since there will be less water used to bulk out the meat. Nowhere is this more starkly apparent than with chickens: all the worst things done to them – from their diet, to the use of antibiotics and chlorine, to the cruelties of unnatural flock crowding and over-fast rearing – are driven by the need to keep prices down. Spending a bit more is good for the planet and good for the chickens – humankind's best friend of all.

CHAPTER 8

PLEASURE AND PAIN

SPICE

'How far Britain has come! The domestic cook many years ago would have been very scared to have anything beyond peppercorns in the house: now Britain has become a country where spices and condiments are imported from across the world. Since the year 2000 our imports have doubled … Britain is on a path of explosion in terms of flavour on the palate.'

Chef Cyrus Todiwala,
The Food
Programme,
2015

Ona sweltering morning in June 1696, Michel Jajolet, Sieur de la Courbe, was working up a hunger beside the Gambia River in West Africa. As representative of the French crown's trading company he had started his inspections at 8 a.m., viewing captured Africans for shipping as slaves to the Caribbean and haggling over the price of each of them. For the new slaves he paid the rulers of Niumi in ivory, lubricating the haggling with French brandy.

The prospect of lunch with a prominent local can only have been welcome. His host, La Belinguere, was alluring and elegant, she spoke several European languages, and the meal was sumptuous. De la Courbe recorded it at some length in his journal.

Among the dishes of elaborately stuffed chickens and 'not at all bad' millet cakes was a complex rice dish. It was heavily seasoned with a fruit he had never seen before: 'green or red, shaped like a cucumber and with a taste resembling that of pepper'. The dish, as Lizzie Collingham says in her history *The Hungry Empire*, was what we now know as the West African staple, jollof rice. The green and red fruits were, of course, chillies – not African at all but traded back to Senegal and Gambia from the Caribbean by the slaving ships.

As Collingham notes, Europeans already knew a form of West African hot spice. The bright red pods of the melegueta plant, a relative of ginger and cardamom, had travelled north much earlier: its black seeds were called 'grains of paradise' in the Middle Ages. Melegueta is still used in West African cuisine, along with another nutmeggy pepper, selim. But the easily grown chilli, born in Central America, is the chief spice in jollof rice today.

SOME LIKE IT HOT

The word 'paradise' appears often in early accounts of piquant foods, like the melegueta: the curious link between 'heat' in food and the pleasure that it gave was recognised by the people of the cold north many centuries ago. Piperine and capsaicin, the active (and similar) chemicals in pepper and in chilli, act on us in a way that nature clearly did not intend. They're meant to frighten us. But – like horror movies – we're thrilled by them.

Professor Barry Smith, philosopher and head of the Centre for the Study of the Senses at the University of London, explained the mechanism in an episode of *The Food Programme* on heat and spice: 'The compounds are part of a plant's chemical defences to deter people and other animals from eating them. Animals just don't have the tendency … we like these spices because they are burning and stinging. No one liked these from birth, we had to acquire a liking for the burning sensation.'

When we eat spiced food it causes us to sweat or feel discomfort – as nature intended. We use the only term we have available in

English – 'too hot'. French, Spanish and Italian use *piquant* or *piccante*, which is a word associated with the pain we feel from an insect's sting. Germans have *scharf* – sharp. Professor Smith says, 'I think we use the word 'heat' because you're actually activating the heat receptors and the pain receptors. The funny thing is, we're bluffing both. We're not getting a real reading of how hot in our mouths is the temperature of the food or a real reading of are we doing damage … It's a kind of taste illusion that we're having with spices.'

The response to the damage warning comes from the trigeminal nerve, which transmits sensory information from nerve receptors to the nose, the mouth and the eyes. Given the data from the action of the spicy compounds, the organs activate their own defences: the eyes produce tears and your nose may start a sneeze to expel any dangerous items.

What is, of course, more neurologically – and psychologically – intriguing is why we do this: why so many of us interpret that damage warning as pleasure. It is a syndrome, even an addiction,

Too darn hot

Piperine and capsaicin and other similar irritants in 'hot' foods are not easily dealt with when you've overdone it. The compounds are not very soluble in water, so flushing with beer does not work – apart from the anaesthetic effect. But the compounds are soluble in fat and sugar, so eating or adding these to the dish – yoghurt or coconut milk will do it – may rescue what appears to be a ruined supper.

based on endorphin highs, something athletes know about. Professor Smith says, 'There's supposed to be a rush people get from putting themselves through that amount of pain [from eating very hot spice]. The body produces opioids that calm you. It's probably akin to the rush you get when you throw yourself out of a plane doing a parachute jump. Thrill seekers will do it and males in a macho way will inspire each other to go hotter and hotter.'

MOTORS OF HISTORY

Many foods lay a fair claim to having shaped the modern world, culturally, geographically and politically. Some of the spices we use to boost bland food or underline subtler flavours with what we call heat or piquancy have displaced peoples, made fortunes, helped build empires. Mass murder and enslavement were a part of that. Of all these world-altering substances, none is quite so unnecessary, so wholly about enjoyment, rather than the practicality of feeding people.

It's often forgotten that when, in the 1490s, Christopher Columbus proposed a voyage west to King Ferdinand V and Queen Isabella I of Castile, it was with the promise of a new route to oriental spices – more alluring than the vague hope of gold. His sponsors were well aware of the enormous fortunes Dutch traders were making trading with the East Indies and his offer was a shorter, less contested route. His failure to come up with anything immediately tradable from the Americas was a huge disappointment to everyone back in Spain. No pepper, no cloves, no nutmeg.

It was this same lure that sent the Elizabethan explorer Henry Hudson off to the Arctic wastes of what would become northern

Canada. He disappeared there in 1511, having found no trade route to the spice lands of 'far Cathay', another term for the East Indies. He left his name on a desolate expanse on the map: Hudson Bay.

But Europe came to discover that the chilli, which Columbus did find in abundance, was exciting. It was cheaper than pepper, for a start. But the plant was just too available, too easily cultivated under glass, even at northerly latitudes, to be of interest to the adventurer-traders to the Americas. Within fifty years of Columbus's return with the plant it had travelled to Goa with the Portuguese traders and took the place it has held at the centre of Asian cuisine ever since. But no one was to make a fortune or fight a war because of chilli peppers.

PEPPER - GREED AND WEALTH

The story of peppercorns is much darker. The berries of vines of the family *Piperaceae* are native to South Asia. They had been used in cooking in the region for at least two millennia when they first made their way to Europe, where demand for pepper would eventually become a catalyst of exploration, conquest and mass murder. The ancient Greeks and the Romans knew and valued pepper, which came to them along lengthy trade routes by sea and land from Asia. By the beginning of the Christian era, pepper was an obsession around the civilised Mediterranean. It changed Roman cuisine as it did Britain's in the Middle Ages: Roman writers mock the greedy wealthy for their overuse of pepper.

The Romans put it in desserts. The second century CE Roman gourmet Apicius has a recipe for honey-cakes of nuts, pine nuts and dates spiced with pepper; the *panforte* that's a Sienese sweet delicacy

The first user of pepper whose name we know was the Egyptian pharaoh Ramses II. As part of the embalming process, a peppercorn was inserted in each of the ruler's nostrils shortly after his death in 1224 BCE.

today uses pepper, too. It is part of pepper's complex appeal that it reacts with acid and with sugar: dressing ripe and sweet strawberries with balsamic vinegar and pepper is a delicious trick. So it is hardly surprising that in early European history pepper was more than a spice. It became a currency, more reliable than coin: there are numerous accounts of rents being fixed in weights of peppercorns – hence the modern term – from the eighth century onwards.

It took another half millennium for the Europeans to develop the marine technology that enabled them to take over the east–west spice routes from the Arab traders, travelling from India, up through the Red Sea and to the Mediterranean. From the fifteenth century, the riches available to wealthy and aspirational Europeans through trading pepper – along with nutmeg, cloves and cinnamon – were to build the first European merchant-empires. This gold rush was the beginning of the conquest and the great looting of South East Asia and India.

The Dutch were the first successful spice traders from Atlantic Europe and in the sixteenth century they dominated. Ships got larger and larger. In July 1599, Admiral Jacob Corneliszoon van Neck arrived back in Holland from Indonesia with a record cargo including 800 tons of pepper and more of other spices: his voyage turned a record 400 per cent profit. The Dutch were doing so well they started trying to purchase extra ships from the English.

News of Corneliszoon's success so excited the English merchants that, a few months later, they set up what was to become the British East India Company. A vicious race for domination of the trade ensued. In 1621 the Dutch rivals – the Vereenigde Oost-Indische Compagnie – took over the Banda Islands in modern Indonesia, burning villages and murdering or enslaving the population. 'No trade without war, and no war without trade,' declared the VOC's governor general, as the company became a financial force to compare with ExxonMobil or Amazon today.

'After the year 1500, there was no pepper to be had at Calicut [in Kerala, India] that was not dried red with blood.'

Voltaire, writing in 1756

The English demand for pepper and other oriental spices was satisfied by piracy: English captains, not so adept at diplomacy and trading as the Dutch, simply seized their ships. The diarist Samuel Pepys, an official of the Navy Office, wrote in 1665 of boarding a captured 'Hollander' ship in London. Its hold was crunchy underfoot with 'the greatest wealth … a man can see in the world,': peppercorns, nutmegs and cloves.

Amitav Ghosh writes that, at one time, a handful of the spices could buy a house. When Pepys was writing, in London a pound of pepper cost 16s 8d, more than three weeks' wages for a labourer and sixty-six times the price of a pound of beef. Cuisine, for the rich at least, had changed radically to incorporate the new exotic ingredients.

Pepys and his wife would sit down, as the food historian Lizzie Collingham writes, to a supper of chicken stewed in stock, salt and pepper, followed by baby pigeon with a ragout flavoured with pepper and other Asian spices. In 1669, Pepys dined with the Duke of Norfolk at Whitehall. He was served what we might now call a salsa, of dry toast and parsley mashed with vinegar, salt and pepper. He called it 'the best universal sauce in the world', to be eaten with 'flesh, fowl or fish'.

But the poor valued peppercorns, too. When the remarkably preserved remains of Henry VIII's warship the *Mary Rose* was raised in 1982, archaeologists found little stashes of pepper in the clothes and the belongings of the seamen, probably for use as currency rather than to spice the shipboard food.

The Dutch VOC's strategy of 'war with trade, trade with war' was enthusiastically adopted by the English spice merchants, proxies for their government. As the seventeenth century wore on, they used superior naval power and bribery of local rulers to squeeze the Dutch and other nations out of the Asian spice trade. The British East India Company (EIC) became a geopolitical force: the first capitalist institution to conquer an entire country – the Indian territories that would eventually be incorporated into the British Empire as the Raj. By the end of the century, the EIC was importing one million pounds worth of pepper annually, and

205

– turning ancient trading routes around – re-exporting it from Britain to the Mediterranean and the Middle East.

Once prices slipped and the novelty wore off, in the eighteenth century, the great lust for the East Asian spices subsided. A bit like salmon today, they were more desirable when only the rich could afford them. Cinnamon, cloves, nutmeg and its outer casing, mace, remained significant in English cooking, especially in cakes and spiced drinks for celebrations. But they were no longer ingredients for which that you might murder a Dutch merchant ship's crew. Or enslave an entire Indonesian island. Pepper became far more affordable as the centuries wore on but it never lost its importance. Even when challenged by chilli, a far cheaper source of sensory heat, pepper has remained central to European cooking and seasoning. Central, too, to many tables: salt and pepper stand at the middle of the altars where we dine. Nonetheless, there's no obvious gastronomic reason for this coupling – neither affects the taste of the other.

Pepper was expensive enough to be taxed in Britain until the mid-nineteenth century and, as with so many other expensive foodstuffs, this attracted opportunists and fraudsters. Cheap ground pepper was sold as DPD, 'dust of pepper dust': it was simply the sweepings of the pepper factory floor, according to the great self-appointed policeman of food adulteration, Friedrich Accum. Writing in 1820, the London-based German chemist told how traders would go to elaborate lengths to fool pepper buyers, making fake peppercorns to insert among the real ones. These were little balls made of linseed, clay and Cayenne pepper (made from dried chilli). Tasked by the tax authorities, Accum examined consignments of pepper and found up to sixteen per cent of the corns to be 'factitious'.

BLACK, GREEN, WHITE AND PINK

Real pepper and its vine come in many forms. There's 'long pepper', which was probably the first to make the journey from India to Europe. It has tiny fruits embedded along a long flower spike. Cubeb pepper, once popular in Britain, comes as individual berries on stems – in Indonesia it is still used to flavour sauces, sweets and cigarettes. There are various African pepper species, including the clove-scented betel, a pepper that's wrapped with lime and other ingredients and chewed. It produces the red spittle you see on the street in some Asian countries.

Black peppercorns are the mature but still unripe berries. They are blanched and then dried in the sun or with artificial heat. White peppercorns, an Indonesian speciality, are fully ripe berries whose fruit layer has been allowed to rot and fall away. Ground, the taste is hardly different; the main use is for cooks who don't want black specks to mar the look of a pale sauce. Green peppercorns are harvested early and then preserved in brine or by dehydration: fresh, their taste is delicate and flowery. Pink peppercorns – different from the more common 'pink pepper' (see below) – are pickled newly ripened red berries.

Szechuan pepper and pink pepper are both misnomers. Adding 'pepper' to a spice's name – as with 'chilli pepper' – has always been a good way of marketing it to innocent Europeans. Both the seeds come from trees and the rind of Szechuan pepper has been prized for centuries; it's a citrus tree and the pungent, tingling, slightly numbing chemicals are characteristic of China's spiciest cuisine. It's thought that the chemical compounds in Szechuan pepper – sanshools – act on different nerve endings at once, confusing our brains. Pink pepper, on the other hand, is a modern invention: the pretty pink berries of a relative of poison oak and ivy, native to the Americas, were first marketed in the 1980s.

Nothing has changed much in the last millennium: pepper is still the world's most traded spice. But the old sources in Kerala and Karnataka, exploited by the British traders in the seventeenth century and onwards, have fallen away as climate change has shrunk the traditional growing area. India today is only the third biggest producer.

Today's spice trade starts in Vietnam, where the country produces three times as much pepper as South India in highly industrialised conditions using only half as much land. Vietnam has history with pepper; French colonialists introduced the plant there back in the seventeenth century, and black and green pepper became important to Vietnamese cuisine. Dishes like *Bò Ham Tiêu Xanh* show the colonial influence – it's a beef and potato stew like you might see in rural France but hugely enhanced by sprigs of fresh green peppercorns.

Vietnam's rise to world pepper leader only dates from the 1980s, after a conscious decision by the communist government to grow crops for export after a long economic slump. In the same period, Vietnam went from almost nowhere to become one of the world's significant coffee producers.

BEYOND PEPPER

Britons have, of course, long had alternatives to pepper.

Ginger was widely available as far back as 1,000 years ago, mostly imported (though it can with care and some warmth be grown here). The root is not so very different from other hot spices. The key compound, gingerol, is a chemical relative of pepper's piperine and capsaicin from chilli. (A tip: wash the root but don't peel it. Most of the flavour is in the skin.) In the Middle Ages, cooks mixed saffron powder with ginger and called it yellow pepper, perhaps because black colours, as in peppercorns, were said to provoke melancholy.

Before pepper became cheap, the ordinary Briton went to the hedgerows and seashore for the spices to lift their cooking. These mostly came from members of the peppery *brassica* family of vegetables, which includes cabbage and broccoli (try eating them raw). First choice among the *brassica* for heat is the root of horseradish – a widespread weed – and of course the seeds of the mustard plant.

A more sophisticated and expensive horseradish, the Japanese *wasabi* is now commercially farmed in southern England by a family of watercress growers, the Olds. Sheila Dillon met them for *The Food Programme* in 2021. Wasabi is a hard-to-tend, palm-like plant, which takes two years to grow in gravel beds with continuous flowing water. It had never before been successfully cultivated in Europe. But the effort is worth it. Its spicy, fruity flavour peaks four to five minutes after cutting into the plant or grating it. 'After twenty minutes it won't taste of anything,' says the farm's operations director Nick Russell. The Olds' wasabi has been a hit, fresh and dried and as a sauce. It won a Guild of Fine Foods Great Taste Award in 2020.

Many British plants have defence mechanisms that turn them into delicacies for humans. Garlic, and its wild cousin, when raw, have the kind of taste-sensor-exciting properties of pepper – sliced, uncooked garlic is used alongside chilli for heat in Thai salads. Alexanders, also known as horse parsley, is the best-known foragers' pepper. The shoots are tasty when young and green, and the tiny dried seeds are spicy with a definite celery flavour. Some fungi also have a piquant flavour: mushroom foragers suggest drying the peppery bolete to use as a spice.

Watercress is also called 'pepper grass' and it gives some definitive buzz to sauces. So can young olive oil. Just as in parts of Spain the new season's oranges are served sliced with the greenest, pepperiest olive oil, bringing out a creamy sweetness, there's an intriguing flavour marriage between oranges and watercress in a salad. Add some salty olives, suggests Niki Segnit in *The Flavour Thesaurus*.

210

PEPPER FROM THE SEA

The delicate seaweed called pepper dulse has long been a traditional food for Atlantic coast communities; more than just a poverty fallback, Irish traditional cooking has always valued it. (One ancient account held that if you were lucky enough to have access to a seaside rock with pepper dulse on it, it was an asset worth three cows.) It is a rich, purple-brown seaweed easily collected at low tide and dried, making a powerful spice with umami-rich, complex flavours. You can cut slices and dry them to make a salty, pungent drinks-time snack.

In a charming episode of *The Food Programme* in 2016, professional forager Mark Williams made a 'wild Scottish curry' for Sheila Dillon, using pepper dulse harvested on a dawn low tide on the Galloway coast. Williams also picked piquant scurvy grass, coriander-flavoured sierra grass, coconutty gorse flower, angelica and hogweed seeds and the chilli-tasting water pepper, more popularly known as arse-smart.

The chef and preserver of Irish recipes Darina Allen writes that care packages sent to Irish migrants from home used to contain little packets of dried dulse, so much was it missed. She suggests stirring it into champ, the classic spring onion, potato, butter and milk mash.

MUSTARD, THE VOYAGER'S SEED

Mustard appears in French recipe books and medical texts as early as the fourteenth century and then in a book by the cooks to the court of the English King Richard II: *The Forme of Cury*. (Cury is an anglicisation of the French verb *cuire*, to cook – in a British-Indian context, though, it comes from the Tamil word *kari* for sauce.) This collection of recipes also mentions olive oil for the first time in an English cookbook and many other spices more exotic than mustard, including nutmeg, ginger, pepper, cinnamon and cardamom.

The mustard plant and its hordes of tiny seeds played a part in the great exploration of North America. The Christian brethren who left the east coast of England for Massachusetts in the seventeenth century carried the fast-growing seeds in their pockets as they travelled, throwing handfuls out along the unmarked ways: 'The idea being it would leave them a trail back to the boat if they got lost or met an adversary. Those mustard trails can still be picked up by satellite today.'

This story was told to *The Food Programme* in 2012 by George Hoyles, one of the pioneers of what now must be seen as the great British mustard revival. That is a tale of how commercial pressure can put traditional farming in jeopardy. Twenty years ago, Britain's mustard growers were hit by low prices from imports and a series of poor harvests. Many were turning away to a more lucrative (and just as yellow) crop: oilseed rape. But, conscious of the noble history of East Anglian mustard farming – and the imminent bicentenary of Britain's best-known mustard brand, Colman's – a group of farmers set up a new mustard growing cooperative in 2007.

Among them was Hoyles, whose family are fourth generation mustard growers. Their grandfathers had made good money from mustard in Lincolnshire, and George was inspired by the need for Colman's – down the road in Norwich – to have at least a bit of its mustard seed sourced locally to earn the 'English Mustard' title on its label. Colman's was founded in the early 1800s but is now owned by the giant food corporation Unilever. Traditionally, mustard is made by soaking the seeds in vinegar and then grinding them into a paste. Colman's adds turmeric for flavour and that distinctive hi-vis yellow.

Plant scientists realised that Colman's had contributed to the story of poor crops among their supplying farmers by abandoning its control on the seed stock it gave to growers after the company was sold. Luckily, some old seeds were found and their DNA analysed, revealing that an important pollinating mustard variety had been lost. Putting that problem right has perhaps saved England's mustard growers. Sadly, in 2018, Unilever announced it was moving most of Colman's production away from Norwich, largely to Germany.

Mustard sauces made on the other side of the country, around Tewkesbury, were famous for centuries as 'the best in England'. They were first sold as dry balls, the horseradish marinated in cider vinegar and then pounded together with mustard seed ground using a cannon ball. Shakespeare refers to a character being 'as thick as Tewkesbury mustard'. An artisanal company in the town has started making it again. It's a different thing from Colman's: a less scary colour and more complex because of the horseradish.

CHILLI, THE USURPER

Of course, once chilli spread across the world in the sixteenth and seventeenth centuries, it was the go-to substitute for pepper: cheap and easy to grow everywhere, given a little sunshine. In the US, the fruit is still usually called the chilli pepper, to distinguish it from the bell pepper used in salads. Neither, of course, has anything to do with the pepper vine.

The British took a long time to adjust to chilli; my mother still won't have it in the house. During her wartime childhood it was used, she believes, to mask meat that had gone off. Prejudice against it has a long history: the great herbalist John Gerard warned sixteenth-century British consumers that while the Spanish used 'Ginie pepper and Indian pepper' to dress their meat, 'it hath in it a malicious quality', adding, 'it killeth dogs'. And, indeed, it is about the only thing that, dropped on the kitchen floor, my dog will not touch.

Perhaps it is this hint of evil and chilli's delivery of hellish pain and confusing pleasure that got the word 'devilled' onto English menus. It describes many recipes where chilli powder is a principal ingredient, rather than a pepper substitute. Shellfish, ham and eggs are devilled. Even today, devilled kidneys, rolled in flour and hot paprika and fried, are still an immovable menu item in the dining rooms of the old gentlemen's clubs of London's St James's.

Cayenne pepper is nowadays simply ground dried hot chillies, though in the seventeenth century it meant a specific type of chilli from what is now French Guiana. It's generally distinctly hotter than standard supermarket 'chilli powder'. Paprika is made from bell peppers, from the same family as chilli – the hotter it is, the more seeds have been used. It was used in Hungary as early as 1569 and paprika remains a defining feature of Hungarian cuisine.

There's speculation about how the fruit got to Central Europe – it's thought that Muslim traders brought it to Budapest from the East. It was an extraordinary and speedy journey. Chilli travelled from Central America with Columbus back to Spain. Then it voyaged to India with the Portuguese traders and then back to Central Europe – all in less than seventy years. It is a testament, if anything, to just how valued the fruit is.

214

There is much competition to produce the hottest chilli of all and a measurement system known as Scoville Heat Units (SHU) was developed to put a rating to contenders. Wilbur Scoville, a US scientist, devised his quasi-scientific tasting regime using trained tasters in 1912. It takes a specific weight of chilli material, dissolves it in alcohol and then works out how much sugared water is needed to dilute the liquid to the point that no piquancy can be detected. The higher SHU, the hotter the chilli. Apart from specially grown hybrids, the world's hottest chillies are the crumpled-looking fruits called Scotch bonnets, which come originally from Jamaica and elsewhere in the Caribbean. They score around 350,000 Scoville heat units; pepper spray used by the police is up to three million SHU.

MYSTERIOUS ORANGE: TURMERIC

Heat can be illustrated with colour, too. Since the Middle Ages, cooks have added spices to indicate warmth and seasonal celebration as much as for their flavour. As well as saffron, cayenne and mustard, red sandalwood was used. But supreme among all of these is the shocking orange-yellow – with a hint of dark green – of turmeric. In its original home, East Asia, this relative of ginger, a similar-looking root, was used as a dye for the robes of Buddhist monks as well as for Ayurvedic medicine and as a foodstuff.

Until recently, colour remained turmeric's chief purpose outside India – more important in modern Britain for putting the yellow into chicken korma and pilau rice than it is as part of the flavour palette. In the nineteenth century, it was most famous as the key ingredient (mixed with flour) in the fake mustard that was common throughout Britain. Arthur Hill Hassall, an anti-adulteration campaigner in the 1850s, reported that on forty-two different occasions he had bought turmeric and wheat flour mix masquerading as mustard sauce. Giving evidence to Parliament, Hill Hassall suggested a factory be set up to produce 'government mustard' from actual mustard seeds, with a guarantee.

Who can say what turmeric tastes like? 'Earthy and bitter', 'almost musky', 'pungent, with a bit of pepperiness' are the vague words cooks use. It adds something: it lifts other, more delicate, flavours in curries. It has been said that 'there is no Indian cooking without *haldi*' and that ninety-five per cent of recipes from the sub-continent use it. In Kashmir, particularly, it is ubiquitous. It is in the first solid foods a baby eats and in the restorative broth served to a mother immediately after childbirth.

There are Indian chefs – like the celebrated Qureshi family of Lucknow – who decry 'tasteless' turmeric's dominance of the nation's cuisine. They say it is not a spice but a medicine – and it is indeed a common first-call antiseptic on wounds. (Try this with powdered turmeric – it certainly stings.) But the Mumbai-born chef and food historian Monisha Bharadwaj sees no contradiction in these roles.

On *The Food Programme*, Bharadwaj cooked a dal with tomatoes, fresh chilli and mustard seed, adding turmeric at the end. She explained her love of the spice: 'Turmeric is often used as a balancing spice. It doesn't have a strong flavour of its own but it tends to bring the other flavours together.' The dal was a big success: 'The whole dish is so complete … brought by the fruitiness, the cleanness of the turmeric.'

Cooking with turmeric goes hand in hand with its medicinal properties: turmeric brings balance and balance brings well-being. 'When you learn to cook [in India], you're always trying to put well-being back into your food because otherwise there's no point in eating,' says Bharadwaj.

TURMERIC UNDER TEST

In the West, turmeric's key ingredient, curcumin, interests mainstream medicine more and more. It has been discussed in relation to issues from allergies to bowel problems and depression. Among the most interesting potential uses is as a supplement for those undergoing treatment for cancer. More research is needed.

Dr Michael Mosley discussed his investigation into turmeric's properties on *The Food Programme* in 2017. He was intrigued by the fact that cancer rates appear to be lower in countries where turmeric is used in quantity. Scientific studies so far, he said, had been 'noisy' – meaning very high doses were used and the results confused. 'We did a study to look at changes in genetic markers because realistically we're not going to be able to follow people for twenty years and see what difference it would make. So we took a reasonably large group of volunteers, around one hundred of them, and we randomly allocated them to either consuming the equivalent of a teaspoon of turmeric or a capsule, or a placebo.'

After six weeks, the researchers found no changes in the placebo group or the one that had eaten the capsule supplements. But the powder group showed substantial and exciting differences. The gene that showed changes is involved in depression and anxiety, cancer and eczema, and asthma – all issues turmeric has previously been said to affect. Are these good or bad changes? More study is needed, say the scientists.

'I took from [the study] that this is a really interesting spice, that appears to have a bioactive role,' says Mosley. He was particularly interested that the capsules of turmeric had no effect. It seems that the spice's bioavailablility – how much actually gets absorbed in the gut – is important. Mosley is convinced to the extent that he is eating the powder. 'I like to scatter it on my eggs … I enjoy it on omelettes and I actually like turmeric latte.'

The latter and things like it have become increasingly available as, since around 2017, turmeric has become the latest in a string of fashionable superfoods. Bee Wilson, in her 2019 book *The Way We Eat Now*, quotes a happy Sainsbury's supermarket executive just launching a turmeric hummus, the latest in a line of cash-in products including tiny 'turmeric shots' in plastic bottles (60ml for £2 in 2021).

**Monisha Bharadwaj's *Haldi Ka Achar*
– fresh turmeric root relish**

Makes about 100g
Preparation: 10 minutes

4 teaspoons fresh turmeric root, scraped and finely chopped
1 teaspoon finely chopped fresh root ginger
1 small fresh green chilli, minced
Fresh lemon/lime juice, to cover the mixture
Salt

Mix the turmeric root, ginger and chilli together, then pour in the lemon or lime juice and season with salt. Transfer to a sterilised, airtight container, seal tightly and shake well to blend. The relish will keep well for up to a week in the fridge.

Recipe from *The Indian Cookery Course* (Kyle Books, 2016) by Monisha Bharadwaj

'Turmeric is a trend that came out of nowhere!' crows the product development manager. But, of course, she is wrong – it came from millennia of practice and experiment in a culture much older than ours.

LOST SPICES

Whether they became unfashionable, too easily available or unavailable because we simply ate them all, many tantalising spices that feature in our ancestors' recipe books are now gone or forgotten. Here's a selection:

Silphium of Cyrene. Possibly the resin of a kind of fennel but no one knows how it tasted. This North African spice was highly treasured by Roman gourmets, perfumiers and lovers. The Emperor Nero is said to have eaten the last example of *silphium*. So the man who played while Rome burned is also responsible for the earliest known example of a plant that was rendered extinct by human greed.

Zeodary. A paler variety of turmeric, the root was used as a peppery spice. It was prescribed by medieval doctors for thirty-six different ailments; the seventeenth-century herbalist Nicholas Culpeper used it with zerumbet (also from the ginger family) to 'expel wind, resist poison, stop fluxes, and the menses, stay vomiting, help the cholic, and kill worms'.

Costus. A strong, peppery herb from the mint family, costus was popular in ancient Rome for flavouring wine and in Norman England for sauces and stews. Grown mainly in northern India and Kashmir, it is now at risk of extinction.

Hyssop. Though it may remind you of laundered towels that have had too much fabric softener, this common herb was valued from biblical times onwards. It was prized for covering up bad smells. British medieval cooks used it with everything from meat stews to garnishes and sweets; it is still around in kitchens in Greece and Israel.

Nard or **Spikenard.** Another classical Roman favourite, where it was used to spice wine. Medieval European cooks used nard for fish and meat. It is mentioned in the Bible. Much faked, it originally comes from a Himalayan honeysuckle.

CHAPTER 9

PEACH OR POISON

TOMATO

'Mrs Beeton told us to boil it for hours and hours, like cabbage. Now, of course, we eat tomato every day of the year … and that has become a problem.'

Food writer and tomato historian Lindsey Bareham, *The Food Programme*, 2012

T he explorers and raiders who went to the Americas in the fifteenth century brought back a basket of exciting fruits and vegetables – what would we be without the chilli, potato and tomato? In their ways, each has changed cuisine and nutrition around the world. After a slow start, now, in the twenty-first century, tomatoes are the world's most popular fruit, central to the food cultures of Europe and the Americas, and popular almost everywhere else. Less so in China, where the average person only eats one tomato for every thirty consumed by a European. But the country is by far the world's largest grower and processor of the fruit, mostly for export. And on that fact hangs a tale.

First, though, a little history. The tomato's journey from intriguing new immigrant to top fruit was slow and circuitous – it took about 450 years. Europeans were thrilled and confused by tomatoes when they arrived from Central America, 1,000 years after the Pueblo and Aztec people started to cultivate them. Were they a poison? The plant was a cousin of deadly nightshade and its acid destroyed the finish on pewter plates. A fruit, a vegetable, a medicine or an aphrodisiac? Or best used as an ornamental plant? For a century, people who were happy to smoke or sniff another New World product, tobacco, spurned tomatoes as toxic. We went on debating what they were for some time: lawsuits were fought over whether the tomato should be taxed as a fruit or vegetable in the nineteenth century.

In 1753, the botanist Carl Linnaeus had decided to put tomatoes in the Solanaceae, the same family as potatoes, aubergines, chillies and deadly nightshade. He called them *Lycopersicum esculentum:* 'edible wolf peach'. The name, perhaps unsurprisingly, did not catch on with consumers. Eventually, the Italians settled on calling what would become a cornerstone of their cuisine 'golden apples', *pomodoro,* alluding to the mythical gardens of the West, the Hesperides, where apples of gold sprang from the blood of a dragon slain by Hercules. But in Northern Europe the tomato languished among botanical oddities for centuries.

One of the first food-loving botanists, John Gerard, put 'Apples of Love' in his *Historie of Plants* (1597), with a drawing of the tomato plant. He wrote that he managed to get seeds from the Mediterranean – though the dragon's blood story was a fable, he decided. Gerard's tomatoes grew in hot horse manure in his garden. But he found them unappealing: cold and watery – 'They yield very little nourishment to the body' and have a 'ranke and stinking savour'. In Spain and Italy, he says, they eat them boiled with pepper, salt and oil to sauce their meat, as the English use mustard.

Suspicion of tomatoes lingered. It is not until the end of the nineteenth century that they begin to appear in popular recipe books, and then rarely for their own sake. The Victorians did eat tomatoes – blanched, peeled and sliced as a delicacy – but *Mrs*

Beeton's Every Day Cookery of 1890 gives only recipes for them fully cooked. She includes five different tomato sauces – a couple for tomatoes stuffed with meat or breadcrumbs and one for them stewed whole in thickened brown gravy.

'Thank goodness someone finally had the courage to eat one of those little golden-red globes and proclaim it for its delicious acidity and wondrous texture.'

James Beard, America's great food writer

In the 1950s, a slew of new, health-conscious cookbooks like Doris Grant's *Dear Housewives* were still suggesting stuffing tomatoes. Although Grant does consider a raw tomato sandwich worth attempting (so long as it is lined with lettuce to stop the bread getting soggy), it wasn't until the 1960s that the British began truly to appreciate tomatoes for themselves, raw. Inspired by the French, we began to eat them with *vinaigrette* in salads, or even as *gazpacho* (though cold soup was for many Britons a step too far). Cooks were

Tinned Tomato Soup Cake

'This is a pleasant cake, which keeps well and puzzles people who ask what kind it is. It can be made in a moderate oven while you are cooking other things, in itself a small but valuable pleasure.' M.F.K. Fisher's recipe for Tomato Soup Cake (butter, bicarbonate of soda, dried fruit, sugar, flour, cinnamon, nutmeg, ginger, cloves and one can of tomato soup) appears in the 'How to Sorrow' chapter of her 1942, *How to Cook a Wolf*, which aimed to help and cheer up wartime Brits suffering under food shortages and rationing. Fisher's wry humour and sarcastic digs at the new social niceties of a time of crisis made the book a classic, though some of the recipes aren't really practicable. Certainly, no one I know who has actually baked her Tomato Soup Cake has been tempted to do it again.

told to dip them in boiled water to remove the skins, which were considered hardly edible.

Long before the English finally got to grips with raw tomatoes, processing and preserving them started a story that would change the shape of twentieth-century food. Key to the epic of the all-conquering tinned tomato was the great Italian migration.

As John Gerard pointed out, the tomato is a fruit that feeds on sunlight and warmth, which not only speeds the crop and increases productivity but boosts sugar levels. But that hardly explains why the Italians took to it with such enthusiasm; poverty and the fact that you could preserve this cheaply grown nutrition is probably the reason. 'It is the life blood of Italian food, some would say of the Italians themselves … a religion whose Holy Trinity is Fresh, Tinned and Concentrate,' writes John Dickie, in his history of the Italians and food, *Delizia!* In Naples he finds more than fifty tomato-based sauces: the city is the 'Tomato's Jerusalem'.

RED SAUCE AND PIZZA

The first recipe for tomato sauce for pasta appears in a Neapolitan cookbook in 1844. Just a decade later, Francesco Cirio introduced tinned and bottled tomatoes, concentrating that delicious sun-got flavour. From being an entirely southern ingredient, the popularity of tomatoes spread across the whole of Italy. Cirio is still one of the biggest tomato corporations today.

Reduced and puréed tomato appeared on flatbreads in poor quarters of Naples from the eighteenth century, and then it was only a short step to pizza. The classic pizza Margherita – tomato purée, mozzarella and basil – was named for Margherita of Savoy, queen of the newly unified Italy in the 1880s: the green, white and red representing the new flag. By then, all of southern Italy was growing and home-preserving tomatoes. And with them came the end, at last, of centuries of endemic famine in the region.

But southern Italy still remained poor. Towards the end of the century, millions of Italian migrants took their tomato addiction with them when they migrated to the United States. At first, there was some resistance to the incoming cuisine: Italians were stereotyped as being 'oily' and 'reeking of garlic' by migrants who had come earlier. But pasta-eating spread during the Great Depression, encouraged by the authorities who saw it as an economic use of wheat flour and approved of the health benefits of the cheap tomato sauce that went with it. Until the Second World War, though, social workers checking Italians for successful integration would continue to note: 'Still eating spaghetti, not yet assimilated.'

Ketchup and kê-tsiap

The Western world's favourite bottled sauce has its origins more than 200 years ago – and during that time it has mutated many times to be the sweet and vinegary red sauce we know today. *Kê-tsiap* is a word in Hokkien Chinese that means fermented fish and China is where British traders first encountered it. They brought the thrillingly pungent sauces home: Lea & Perrins Worcestershire sauce, based on fermented anchovies and spice, is a direct descendant. By the eighteenth century, British and American cooks were buying and making all sorts of preserved spicy sauces, calling them Catsup or Ketchup, with bases of mushrooms, nuts and shellfish.

Charlotte Mason's *The Lady's Assistant* of 1787 has intriguing recipes for walnut ketchup, involving the nuts, anchovies, garlic, vinegar, mace, cloves and 'Jamaica pepper'. The first recipe including tomatoes appears in 1817: it caught on. By the early twentieth century, Henry J. Heinz (who started out in horseradish sauce) was shipping twelve million bottles of tomato ketchup a year. Its success, rivals have sourly pointed out, may be mainly due to the fantastic amount of sugar in it: twenty-five per cent of the product. One tablespoon has more than you'd find in the average chocolate chip cookie. But then so does Ketjap Manis, the gorgeous, sweet, spicy soy sauce ubiquitous in Indonesia – another place to which kê-tsiap migrated.

227

PIZZA COVERS THE PLANET

The simple Neapolitan flatbread with tomato sauce and various cheeses on top was unstoppable. In the 1950s, 'pizza parlors' started to spread from Italian neighbourhoods to other areas of America's big cities. At the same time, society – and the food industry – discovered teenagers as a consumer group: fast food was as ideally suited to a casual, car-driving, outdoor society as it was for a TV dinner on the sofa. Chefs worked out that pizza dough could be spread on pans in the mornings and left in the fridge during the day, ready for tomato sauce and toppings and a blast in a hot oven when the evening trade began. This drastically speeded up serving time.

By 1956, the pizza had overtaken 'ham and eggs' and the hotdog as the top meal in American restaurants and homes, and it has never looked back. In 1958, Pizza Hut started making the dough and sauce centrally: all the company's franchisees then had to do was bake the flatbread with the toppings. Pizza Hut opened in the UK in the late 1970s (fifteen years after Britain's first pizza chain, Pizza Express, opened in London's Soho) and it now has outlets in more than ninety countries. In 2001, it supplied the first pizza outside Earth: a tomato, mozzarella and salami 'six-incher', which a Russian astronaut ate on the International Space Station.

With the hamburger, pizza is now the most popular fast food the world has known. Norwegians eat the most pizza per head, an average nine each per year, most of them frozen and baked at home. China is the fastest-growing pizza market, while forty per cent of Americans eat a pizza every week, bought from 77,000 different outlets. Most Northern Europeans eat more pizza per head than do Italians.

Tinned tomatoes

The experts' advice is to buy tins of whole tomatoes – not chopped – and make sure you use a good brand. The chef Mary Contini of legendary Edinburgh grocers Valvona & Crolla suggests being choosy about the brands of tinned tomatoes and tomato paste you buy. The brands Mutti and Cirio are reliable. (Mary looks for labels boasting the famous San Marzano cooking tomatoes of southern Italy.) The difference between economy tinned tomatoes and more expensive ones is massive, when tried side by side, though labelling frauds are said to affect fifty per cent of Italian processed tomato exports. Check labels for any additives.

The science of the obsession is easy enough: in a classic pizza, grilled cheese and rich tomato sauce are both high in taste-enhancing glutamate and, mixed together, appeal to human taste receptors even more than they do separately. The other key factor is, of course, in the manufacturing: pizza is cheap to make and easy to transport, which was the point of it when it was first delivered to Naples's slum-dwellers without their own ovens. The mark-up on the cost of ingredients, depending of course on toppings, of a fresh pizza is somewhere around 700 per cent.

GROWN FOR FLAVOUR?

The tomato-growers' challenge is chiefly about replicating the effect of the sun's heat to raise the tomato's sugar content and get the fruit to be the right colour at the right moment on the supermarket shelf. Another problem is transport and storage: refrigeration can kill the taste. Industrial production is the source of much that's wrong with tomatoes today, from a tomato-lover's point of view and for those of us worried about the climate crisis.

'In 1999, the big buzz word for tomatoes was "grown for flavour" – which begs the question, what else do you grow them for?' asks the writer Lindsey Bareham, whose *Big Red Book of Tomatoes*, published that year, was a surprise bestseller. In it, she pointed out acerbically that most of the tricks used by tomato distributors and sellers were for the eye, not the tongue.

Tomatoes must be refrigerated to be trucked from Southern Europe, the origin of most tomatoes eaten in Britain outside the summer and autumn. But tomatoes kept in the fridge will taste acidic and watery; they need a few days at room temperature, out of plastic, to recover. 'Vine-ripened' is a scam, reckons Bareham. Laying out a branch of vine-attached tomatoes in a tray is just 'a licence to attract exorbitant prices'.

Science is the parent of the modern supermarket tomato. Heinz was the first large tomato processor outside Italy to begin tweaking the genetics of the fruit for profit, starting an experimental growing and breeding programme for its own farms in the 1930s. Glasshouse tomatoes take between forty and sixty days from flowering to get to full ripening but they can be picked earlier and treated in the warehouse or in trucks with ethylene gas, which promotes the production of lycopene, a natural carotenoid that gives the tomato its red colour. This is what has happened when you pick up a perfect, fire-engine red tomato that is still as hard as an apple: inside it will still be immature and taste of very little. Since the trip from its greenhouse in Holland or Spain to the shelves in Britain can take four days or more, this is a significant advantage for the producers.

In 1994, the Flavr-Savr was born. It was the first foodstuff genetically modified in a laboratory – rather than by breeding – to obtain a licence for human consumption. Compared with today's GM creations, it was a simple thing: one tomato gene was isolated and reversed to slow the ripening process. But one key thing the Flavr-Savr lacked was, despite the name, flavour. No one liked it and it was soon withdrawn, though it remains famous worldwide because of its test-tube parentage.

Genetic modification and gene editing has continued, producing experimental tomatoes that address ripening speed and tolerance to frost and diseases. One has been produced in India that can stay 'ripe' and visually perfect at room temperature for forty-five days. Tomatoes, tweaked and fiddled with, seem certain to continue to evolve in a way that may benefit business far more than the tomato-lover.

POLITICISING THE TOMATO

Across the European Union, genetically modified (GM) crops must get through tough testing and regulation. Currently the only commercially grown GM crop in the EU is one variety of maize, altered to protect it from insects and used for animal feed – though several European countries have banned it. But Britain's exit, under the Boris Johnson deal, may eventually allow such crops and foods made from genetically modified plants and animals to be sold in Britain for human use. In early 2022, new rules made it easier to grow GM crops for research and use techniques like gene editing in the laboratory. The British government has stated that it wants to move on from 'outdated' EU regulations on GM and other new technology, though the Scottish government has demanded that there should be no change to the regulations.

A trade deal with the US, where some GM foods are permitted – including soya and tomato sauce – would pose a problem, as Dan Saladino explained on *The Food Programme* episode 'Brexit: The Tomato's Story' in 2019. But some influential Conservative MPs in rural areas think British farmers should be allowed to use GM technology to compete in world markets where rules on food are less stringent.

In Britain, work continues on ever more complex genetic alteration of tomatoes in the hope of a regulatory change. In 2012, the John Innes Centre of plant research in Norwich announced an astonishing purple tomato, deep-midnight velvet coloured, produced by inserting a gene from the snapdragon flower. This allowed more anthocyanin to develop. From that came not only the impressive purple colour but also a longer shelf-life and less danger of 'grey mould' evolving. There was also an additional potential health benefit and it was this that the scientists trumpeted: anthocyanin has been shown to slow cancer growth in mice and has anti-inflammatory effects. It is why people are advised to eat naturally purple fruit like blueberries.

The John Innes tomato went to Canada for bulk growing because of the European regulations on genetic modification, and 2,000 litres of purple tomato juice for drinking came back – legally, because all seeds had been removed. A reviewer for *The Food Programme* cruelly dismissed the offer of a Bloody Mary made with it. Far from purple, the purée was a 'grey sludge'.

Ten years on, the purple tomato still has not been licensed commercially but the leader of the team that invented it, Dr Cathie Martin, has continued editing the genes of tomatoes. Her latest invention is a tomato tweaked to produce L-Dopa, a drug needed by Parkinson's disease sufferers. A spokesman for the John Innes Centre explains that in poorer countries, a day's dose of L-Dopa can cost $2 but growing the tomato can get round that, and also offer all the normal benefits of a tomato crop.

There's a popular myth that GM has been used to produce square tomatoes, easier to pack and store. It has not. In fact, producing a square tomato is something you can do quite easily in your own home. If you grow tomato plants, you can make a clear plastic box with a sheet of semi-rigid filing plastic, scissors and tape. Suspend the box so it hangs around a likely looking ripening tomato and at the right moment you will get a square one. They're not much easier to store – but you can play Jenga with them.

THE CONSEQUENCES OF CHEAP TOMATOES

Ever since people in Britain and Northern Europe started demanding fresh tomatoes all year round, terrible stories have emerged from Southern Europe about the conditions of workers where tomatoes are farmed. This does not seem to have made any impact on our appetite for them. In 2018, sixteen immigrant farmhands died in forty-eight hours in the over-crowded trucks that transported them to the tomato fields in southern Italy. Today, most of the workers are recent arrivals from North Africa: as illegal migrants they have no labour rights or recourse to the law, fearful of being deported.

That same year, the United Nations' special rapporteur on slavery said that 400,000 agricultural workers were at risk of being exploited and almost 100,000 were forced to live in inhumane conditions. The Italian government at the time was accused of tolerating the situation because it viewed any attempt to make conditions better as merely providing an incentive for more migrants to cross the Mediterranean.

An investigation by the *Guardian* newspaper in 2019 found slum camps across southern Italy, where migrant agricultural workers lived in unsafe and unhealthy conditions to serve the farms. Rapes and deaths are frighteningly common. According to the season, they picked tomatoes, peppers, grapes and potatoes. The work was paid by weight picked – €3.59 for 300kg tomatoes. It was hard to

234

Tomatoes and the climate crisis

A kilo of basic commercial tomatoes grown with fossil-fuel heating can produce as much as 10kg in greenhouse gas emissions – as much as is produced by producing a kilo of ham. If you are serious about the climate crisis, out-of-season tomatoes are clearly a no-no: there are clever ways of heating greenhouses, through solar energy and use of waste heat, but only premium product growers can afford them. A tomato salad grown locally, using only a sun-heated greenhouse, is judged to produce 8g of g CO_2e; from a fossil-fuel heated greenhouse it involves 626g of g CO_2e. Transport of the tomatoes to Britain from Holland or Spain, which is generally how we get them for eight months of the year, would add another five per cent or so.

make more than €30 a day, out of which employers would deduct fees for transport to the fields, and more if an accident meant a trip to hospital. Organised crime has been busy on the farms since the 1990s. There are many disturbing stories.

In 2018, one tomato farm in Puglia, specialising in a more valuable biodynamic crop, promised a decent wage and good conditions to its workers but within a few days all its tractors had been stolen. The core problem, according to Italian analysts, is the price Northern European retailers are prepared to pay. 'The link between the cost of work and the price of the product has broken,' wrote Yvan Sagnet, a former tomato picker, originally from Cameroon, and Leonardo Palmisano, in a book called *Ghetto Italia* published in 2017. In winter 2019, a kilo of tomatoes paid a farmer in Puglia 7.5 cents. In a supermarket in Britain they would sell for one hundred times as much.

Today, Italy is in the world's top ten of tomato exporting countries. Strangely enough, it also came in at number seventeen in the lists of top tomato *importing* countries. That is because China grows tomatoes more cheaply than even the illegal labour-driven parts of Southern Europe. As with olive oil (see the chapter on fats and oils), EU labelling legislation permits bulk products to be imported and repackaged in Italy and labelled 'Produce of Italy'. Some reprocessing has to be done, but in the case of tomato paste, that may only mean the addition of salt and water.

Italians protest that frauds around their foodstuffs are not a sign of lax regulations in Italy, merely an indicator of the value of 'Made in Italy' as a label worldwide: it is hardly their fault fraudsters are attracted. That may be so but China, now the world's biggest producer of tomato paste, says most of its production is exported to Europe. Other than in an Asian foods shop, no supermarket tomato paste brand in Britain carries a 'Produce of China' label.

With governments uninterested or unwilling to intervene, as always it seems that the onus to stop all this falls on the bargain-hunting consumer: if we insist on food at prices that are a fraction, in real terms, of what our grandparents would have paid, we are liable to get products where corners have been cut – in the ugliest of ways. The price paid to tomato growers in the UK did not rise at all in the first decade of this century while input costs from fertiliser to labour all rose. At the same time, some British supermarkets did no-price contracts with desperate farmers: the supermarket sold them for what it decided, even at buy-one-get-one-free. If the tomatoes went off or could not be sold, farmers paid the cost of disposal.

With Chinese tomatoes there is even more to be concerned about. Most of the fresh tomatoes China produces annually come from state farms in the western province of Xinjiang, where the use of forced labour by detained members of the Uighur minority group has been reported. In early 2021, the United States government and some Japanese corporations stopped imports of tomato-based products from Xinjiang for this reason. China denied the accusation. Chinese tomato paste goes to the UK and the EU, mainly Italy: no country here has, as of late 2021, taken any action.

So, as a tomato lover, what do you do? From slavery to fossil fuel-derived fertiliser and genetic modification there's quite a bit to worry about. But some things are getting better. Hydroponic cultivation is becoming a reality. A long way from the classic greenhouse, free of soil and pests, this technique – popularised by industrial cannabis farmers – means tomatoes can be grown with minimal water use, provided with only the nutrients they actually need, though healthy trace elements that normally pass from the soil to the tomato have to be added to the feed.

Indoor 'vertical farms' are another alternative to conventional tomato growing. Using artificial lights, recycling water and nutrients, produce is grown on towers of shelves. This reduces land use and pollution and is already common in crowded cities in the United States, Japan, China and Singapore. Britain's first vertical farm, the size of one hundred tennis courts, will open near Scunthorpe in 2022. Positioning tomato farms where waste heat from industry or power generation is available is one other energy-saving notion that is becoming mainstream.

Chef Jeremy Lee's Jug of Perfect Bloody Mary

Jeremy uses Pago tomato juice at his Soho restaurant Quo Vadis. 'Worth considering is the rather natty way they have in New Orleans of serving wild and wonderful pickled vegetables atop the Bloody Mary, such as okra and green beans.'

Serves 5–6

300ml vodka (I use Stolichnaya)
150ml lemon juice, just squeezed
1 litre tomato juice
20ml Worcestershire sauce
18 splashes Tabasco
2 heaped tsp celery salt
1 tsp freshly ground black pepper
Celery stalks, lemon slices and ice to serve

It isn't true – though often repeated – that tomatoes today have only one-tenth of the nutrients they had fifty years ago. (It has been pointed out by the BBC that we don't actually have the data from 1970 to show that.). In fact, with the rise of heirloom tomato cultivation, and more attention given to growing slowly, there's probably a better choice of tomatoes for consumers in 2022 than there has ever been.

'It's an absolute parade of tomatoes: little round ones, dark green ones, spiked ones, ribbed ones,' reported Marlene Spieler on *The Food Programme*, visiting a grower in the Isle of Wight epicentre of the growing heirloom tomato trade. She saw beautiful, complex-tasting tomatoes like the giant Jack Hawkins, cherry-sized Sunshine Yellow, the Red and Green Zebra (striped orange, red, white and green), the long-flavoured local Island Beauty and the super-tasty Angel and Piccolo. Spieler loves an Italian dessert tomato dish: you dip the cherry tomato in hot caramel, like a toffee apple. The potted advice – for flavour, for health, for the workforce and for the planet – is to buy small tomatoes (generally they have more flavour), grown in-season and be very wary of bargains. They may have cost people far away and poorer than us more than we know.

CHAPTER 10

A WORLD IN A SEED

RICE

'Apart from being nutritious, rice produces much larger
crops per seed sown in the ground than any other
cereal. It tastes good, it has a good texture, it stores
well, it's easy to transport, it doesn't require
elaborate preparation: it's just the perfect
food. For half of the human race rice is
not just a major article of daily food,
it is part of a way of life.'

Roger Owen, author with
Sri Owen of *The Rice
Book* speaking
on *The Food
Programme*,
2008

K

hamphou Pholsri is busy cooking rice for the first sitting at Passorn Thai, the Thai restaurant where he has been the chef for ten years. He tips one-and-a-half kilos of premium grade jasmine rice – *khao hom mali* – into a huge bowl and puts it under the cold tap. Pholsri swirls the stubby white grains, releasing a cloud of milky starch and dust, pours the water off and does the same again. After a third rinsing the rice is ready for the cooker. Into the pan it goes with more cold water, the level just a knuckle and a half above the top of the rice. Sixty years old, Pholsri has done this ten thousand times or more, ever since childhood. He knows exactly how much water *khao hom* needs, how much this particular brand needs, to the last drop.

This is important because rice should never be cooked with more water than the minimum necessary. Chef Pholsri, like so many Asian cooks, is amazed by what the British think is the correct way to do the job. We don't rinse it, we stir it while it cooks and then, worst of all, we drain away the excess water when it is done.

240

'You're losing the taste, the texture will be wrong and there will be no lovely smell!' he laughs. Rice has a smell? It is an alien notion to a people who think that, as with pasta, taste and odour is something that you add later on, with sauce. Yet, in Asia, to the world's primary rice eaters, the scent of the grain is prized: *basmati* is not named for a place but a word meaning 'fragrant'. Thai *khao hom mali* is called 'jasmine rice' because that's how it smells.

Khamphou Pholsri grew up in Isan, north-eastern Thailand, working in his family's rice paddy when he had time off school. His face broadens into a smile as he remembers the rituals that marked his childhood when he ate rice – three times every day, as most Thais do.

'Last thing at night my mother would put the *khao niao* [sticky or glutinous rice] on to soak. In the morning, she would be up at dawn, steaming it. She would press it into lumps, brush them with egg and bake it so it had a golden varnish. We'd eat that for

'Gin khao ruk yang?' is a common Thai greeting, meaning something like "How are you doing?" Its literal translation is "Have you eaten rice yet?"'

breakfast and then take some of the cakes to school to eat with spicy fish or *som tam* [northern Thailand's staple green papaya, tomato and chilli salad]. In the evening, we'd come home and eat *khao suay*, which is steamed rice – or, literally, beautiful rice – with some meat sauce.'

Some four billion people on the planet eat rice every day. They are dependent on the grain for the bulk of their nutrition. In many countries in Asia the words for 'food' or 'eat' and 'rice' are the same. We eat far less rice in Western Europe – only 5–7.5kg per head a year, compared to 250kg or more in some Asian countries. All over Asia, rice has long been a symbol of wealth, fertility and godliness: the name of the Buddha's father, Suddhodana, means 'grower of pure rice'.

So, when rice crops fail or are scarce, the devastation is more than economic: it is a crisis of the body and the spirit. Already this century (see page 252) one rice famine has befallen the poor of East Asia and it seems likely more are to come. Rice and water go together so climate crisis challenges are likely to affect this crop, which involves more than one billion people in its production, more than any other on the planet.

On Northern European tables, there are alternative carbo-hydrates: rice has only become a staple in our meals in the last century. Now the British and the Italians eat more rice than other Europeans, and both countries have intriguing historic and cultural reasons for their rice habits. The northern Italians grow rice, of course, and have done so since the late Middle Ages. The British don't grow rice – yet – and the fact that we now eat more than Italy does is all to do with immigrants bringing their rice needs and their cuisine from Asia.

MAGICAL SEED

Rice is an extraordinary grain. It is, by origin, a dry land plant, like other grasses, and can be farmed like them in fields with a little irrigation. But long ago, rice evolved a strategic alternative: it could exploit the monsoon season, and the huge volumes of water that come with it, by storing air in bubbles along the surfaces of the plant that sit below the water, providing an alternative oxygen supply. If torn from the ground in severe floods, rice stalks will float until they can put down new roots. Most extraordinary of all, rice growing in this risky medium is far more productive than on dry land, a crucial advantage that humans have been leveraging for millennia.

Thus, the agriculture of rice is like no other food crop, using dyke systems to flood the land and force the crop. These can simultaneously be used to farm fish and then in the dry season animals can be fed on the rice stubble. Two or three crops a season can be had from the paddy; at the right moment, rice stalks in the paddy fields will grow as much as 25cm each day to keep the heads above water. It is an amazingly efficient system in which the land is

Rice vs wheat – who wins?

Rice is the most productive of all the grasses humans grow and harvest. One grain can become a plant that will give 1,000–1,500 more seeds, 23–35g rice. A wheat plant will produce only some 120 kernels, about 5g. But wheat is more nutritious. Gram for gram, wheat flour has nearly three times the calories of cooked rice, brown or white, and more vitamins and minerals.

in use twelve months a year, its fertility constantly refreshed by the debris in the water and then by animals who graze the paddy in the dry season.

SLOWLY IN EUROPE

Rice's key attribute, as a crop, is that you can get so much from it – more than any other grain. The fifteenth-century Duke of Milan Galeazzo Maria Sforza is credited with spreading its cultivation in the well-watered plains of Lombardy in northern Italy. He sent sacks of rice to other landowners, promising them that one grain would reproduce twelve-fold without much attention. Sforza goes down in history as a corrupt and cruel ruler but from his passion begins one of the world's great rice cuisines. By the end of his life, rice was being grown along the marshy plain of the River Po, thanks to a drainage and canal system in whose design Leonardo da Vinci played a part.

The Romans were eating rice, probably imported from the Middle East, as an expensive remedy to settle the stomach 2,000 years ago. Though often ignored in books about the spread of civilisation via food, people in West Africa living in the floodplains of the upper Niger river delta independently domesticated the African 'red' rice species 2,000–3,000 years ago. It became an important food crop across the region, farmed in complex paddy field systems not unlike those in Asia.

Legend has always had it that the thirteenth-century explorer Marco Polo brought rice flour noodles back from China to Italy. Impressed – who would not be? – the Italians used wheat flour to replicate the dried strings and have been grateful ever since. (Although it's now thought the Polo story was put about through an advert for a Canadian food company trying to popularise spaghetti there in the 1920s.)

Though the Chinese first domesticated rice, and developed the techniques to mill it, the farming of it came west well before Marco Polo. In Moorish Spain, rice was cultivated by the Arab people from North Africa 800 years before Duke Sforza introduced it to northern Italy. It arrived in medieval northern Italy through Venice and its trade routes.

Outside China and West Africa, for centuries rice was an expensive delicacy, usually eaten as a sweet dish. 'White rice with

Kedgeree – an imperial fusion

The British had one notable savoury rice dish, popular from the late eighteenth century on. It is a perfect summary of imperial-era influences and nostalgia, not least in the confusion over its actual roots. We call it kedgeree and today it means rice cooked with onion, curry spices and butter, topped with flakes of smoked fish, usually haddock, slices of egg, parsley and turmeric. It was a crowd pleaser at Victorian tables; it reminded returning soldiers and administrators of their Indian days, it is said.

But its etymology is vague. People eat fish with rice in coastal India but the Gujarati *kidgri* – the closest word – is made from mung beans or lentils, fried onion, ginger and rice. Nonetheless, this first Anglo-Indian fusion dish was a favourite of Victorians including Queen Victoria and Florence Nightingale, and kedgeree deserves more attention today.

melted sheep's butter and white sugar is not of this world,' wrote the eighth-century Arab philologist al-Asma'i. Rich people in Tudor England ate rice, probably imported from Spain, baked with cream, honey, dried fruit, nuts and the exotic spices like cinnamon and clove then appearing in London. Rice puddings remained so through to the twentieth century, though sometimes the rice was boiled or baked in milk and sugar and served with very little else. As a result, generations of British schoolchildren learned to hate the dish – the skin on the top of the congealed rice being particularly revolting. The author of *Winnie the Pooh*, A.A. Milne, wrote a poem deploring it.

In her 1975 classic, *English Food*, Jane Grigson includes no rice dishes other than rice pudding. She cooks it in creamy milk with sugar, just as al-Asma'i liked it 1,200 years earlier. Bad rice pudding can make you queasy, she admits, with her own school dinner memories still fresh. To avoid that, 'A rice pudding must be flavoured with a vanilla pod or cinnamon stick, it must be cooked long and slowly, it must be eaten with plenty of double cream. Like so many English dishes it has been wrecked by meanness and lack of thought.'

Rice did not become a staple in Northern European cooking until the twentieth century. In the late Middle Ages, it was so valuable, in Britain, that it was kept in a locked cupboard, with the spices, and referred to in the singular: 'Take rice and wash them clean'. As with so many luxury foods, it had its enemies, on moral grounds. One Elizabethan writer was convinced boiled, spiced rice 'will provouke into venerie' – inspire lust. Others at the time prescribed it for 'improving the flow of a mother's milk'. The eighteenth-century gastronome Jean Anthelme de Brillat-Savarin thought that rice eaters were soft and cowardly, holding up the Indians, whom he considered rather easily conquered, as an example. Before modern

'A very pretty dish': Carolina Snow Balls

In Britain, wealthy people in the late eighteenth century enjoyed 'Carolina rice pudding' and 'Carolina Snow Balls' – the long-grained, fluffy grain being judged far superior to any produced in Europe:

'Take half a pound of rice, wash it clean, divide it into six parts; take six apples, pare them and scoop out the cores, in which place put a little lemon-peel shred very fine; then have ready some thin cloths to tie the balls in; put the rice in the cloth, and lay the apple on it; tie them up close, put them into cold water, and when the water boils they will take an hour and a quarter boiling; be very careful how you turn them into the dish that you do not break the rice, and they will look as white as snow, and make a very pretty dish.

'The sauce is, to this quantity, a quarter of a pound of fresh butter melted thick, a glass of white wine, a little nutmeg, and beaten cinnamon, made very sweet with sugar; boil all up together, and pour it into a basin, and send it to table.'

From Hannah Glasse's *The Art of Cookery* (1796)

times, apart from the puddings for wealthy gourmands, rice was mainly used on ships or in the military as a foodstuff because it was portable and stored well.

Even so, the British were busy in the rice business long before it became a conventional kitchen cupboard item. Enslaved people, taken to North America from West Africa, brought with them rice seed and the knowledge to farm it. Soon the grain grown by slaves in the American state of South Carolina became an important means of feeding enslaved workforces and by the end of the seventeenth century, the state was a centre of rice farming.

As in West Africa and across Asia, low-lying fields were converted to paddy with drainage channels. By the mid-eighteenth century, Carolina rice – probably a hybrid of the African variety and seeds brought from Asia by East India Company trading ships – was being exported in bulk to the slave colonies in the Caribbean and to Northern Europe. Rice and beans, jambalaya and, in Jamaica, rice and peas (exported dried from Britain) remain defining dishes for the people of the southern United States and in the Caribbean: comfort food that arises from exploitation and misery.

The demands on the enslaved people working in the rice plantations and processing seem incredibly harsh today. The

'It's as much part of my life as my dad's Fela Kuti records or going to wedding parties as a young Nigerian boy...'

On *The Food Programme* in 2018, musician Tayo Popoola savoured jollof rice, one of many culture-defining rice dishes that can be found across the world.

threshing and winnowing – the separation of the grain of rice from its husk – was done by hand. Men were assigned 30kg of rice a day to pound in a pestle and mortar; women were given 20kg. Machines to husk and polish – rub away the inner surface of the rice grain – did not appear until the 1860s.

Meanwhile, the sophisticated rice farming systems of the West African river valleys all but disappeared in the same period, probably because of labour shortages caused by enslavement. Between the seventeenth century and the late nineteenth century more than fourteen million West Africans were taken by force to the European colonies in the Americas.

THE RISE OF SUSHI

You could start a good argument over which nation has taken rice cuisine to its highest forms. For the British, without home-grown rice, the last seventy years has seen an extraordinary procession of amazing immigrant rice dishes, all the more flabbergasting to a nation whose most sophisticated rice cuisine – beyond that ubiquitous pudding – was using it to soak up sauce. In quick succession, British food absorbed flavoured rice in the form of pilau, paella and risotto, and learned to distinguish between them. Then came rice's most exotic, and expensive, iteration ever: sushi. It is only twenty years since a British entrepreneur, Simon Woodroffe, brought the Japanese conveyor belt system, originally invented in 1958, for serving the ancient street food to British cities. At the time, a Yo! Sushi marketing executive told me she did not believe her Geordie, working-class parents would ever sit down to a plate of raw fish and vinegary rice eaten with bamboo sticks; nonetheless, forms of sushi were on sale, pre-prepared, in all British supermarkets' chiller cabinets by 2010. And her parents came round to it, too.

Sushi as we know it – a lozenge of rice with a strip of fresh fish on top – began as a street food in mid-nineteenth century Tokyo: a snack for busy people. The vinegar taint in the rice remembers the use of acid from fermentation to keep bacteria down in the days before refrigeration. The fish for *sushi-meshi* would all have come from the bay of Tokyo. The food was surprising enough to be mentioned by foreign visitors in accounts from the era: 'a roll of cold rice with fish, sea-weed, or some other flavoring,' as an 1893 book puts it.

Today's complex sushi confections have evolved quite a way. The California roll was invented by a Japanese-Canadian chef in the 1970s, though the idea of a sushi roll with rice on the outside – *uramakizushi* – has a longer history. Most surprising of all to the nineteenth-century Japanese would be the ubiquity of raw salmon, not a Japanese favourite because of the high likelihood of the wild fish being worm-riddled. (All sushi salmon eaten now is farmed with the use of pesticides.) They might also have an issue with the sushi in the chiller cabinets of British supermarkets: cooked fish, mayonnaise and not nearly fresh enough. Not just any rice will do, either: short-grain sticky *shari* rice, perfectly cooked, is crucial. In Thailand, a great fuser of cuisines, American-style pizza restaurants serve sushi-pizza, with the fish and rice rolls plonked on top of a cheese and tomato base.

Japanese eating is elegant and adorned with etiquette in a way we may never be able to match. But there are some simple rules for a sushi restaurant, conveyor belt or otherwise:

- Nigiri, the rice fingers with a slice of fish or meat moulded to the top, come in pairs because one is bad luck.
- It is perfectly acceptable to eat nigiri with your fingers – they were designed as snack food for hurrying city folk. Chopsticks should be used for sashimi, the rice-less slices of fish.
- Nigiri should be eaten fish-side down, and only the fish dipped in the soy sauce. Wasabi will have been added already: you need it for *maki* and *futomaki*, the rolls, not for nigiri.
- The pickled ginger is for palate-cleansing between different sushi mouthfuls. Don't eat it with them.
- Sake wine is made from rice, so in Japan it is thought unsuitable for drinking with rice-based food. Try green tea instead.
- If you use a napkin, it is polite to leave it folded afterwards.

Many ways with rice

Rice is a staple in one hundred different countries and we all do very different things with it – from main dishes to rice wines, from the delicate, diaphanous rice-paper-wrapped spring rolls of Vietnam to Hoppin' John, the sturdy plantation-era dish of black-eyed peas, rice and collard greens that is native to the southern US.

Sri Owen's *The Rice Book* is a much-loved bible, first published in 1993, of rice lore, history and cooking. She lists hundreds of wildly different ways with the grain. One of her own family recipes – she is of Indonesian origin – is a black rice ice cream made of liquidised glutinous (sticky) black rice, coconut milk, cinnamon and salt: it is delicious.

THE 2008 CRISIS

It is clearly dangerous to depend on one food crop, as the people who suffered potato famines in Northern Europe in the nineteenth century discovered. But some people don't have much choice. The poorer a rice-dependent nation is, the more rice its people eat, per head. The top ten rice-eating countries are a table of some of the world's most disadvantaged and strife-torn countries, beginning with Cambodia, Bangladesh and Laos. People in all of them eat over 220kg per head or more every year. Sierra Leone, Guinea and Madagascar come closely behind. Famous rice eaters like the Chinese eat rather less, at 125kg, the Thais, 175kg.

The first major food crisis of the twenty-first century was remarkable because it appears – trade experts are still debating it – not to have been caused by a climate issue, disease or war, as most such disasters are, but by economists and the inflexible diktats of neo-liberal ideology. It was dire. In the twelve months to June 2008, the price of rice increased by 170 per cent. The price of wheat more than doubled, too – but wheat eaters, unlike rice consumers, generally have greater access to alternatives. Rice traders in Britain complained, as Sheila Dillon reported for *The Food Programme* later in 2008, that there were all sorts of rice that were currently unobtainable – they forecast empty spaces on British supermarket shelves.

We could live with this but for the half of the world where rice is the main source of energy and a way of life, these price rises were disastrous. Low prices earlier in the 2000s onwards had meant that stocks of rice were low, as poorer producing countries sold off emergency stock as they sought to keep export earnings up, and there was little cushion for the rice-users. By mid-2008, the biggest rice producers, including Thailand, Vietnam and India, had, in panic, stopped any exports in order to protect their own consumers. In Cambodia and Bangladesh, where more rice is eaten per head than anywhere else, the price of rice suddenly rose – up fifty per cent in six months.

This disaster came with a terrible irony for those living there: harvests had recently been good. Both countries had produced a surplus of rice over the previous two years – and exported them, at the behest of the international financial aid bodies, to their richer neighbours, like Vietnam, India and Thailand. I visited two families in rural Cambodia in 2008 to find out what this meant to their lives. Their hunger was evident: none of the adults in fifty-two-year-old Lai Phon's house were going to get enough rice to satisfy their needs that night, once again. Rice in Cambodia provides sixty per cent of ordinary people's nutrition: when there is a shortage, millions go hungry.

Lai Phon cooked the rice they had for the nine family members while we talked. She and her neighbours had taken their children out of school because they needed to work to help the family budget. Why did she think the price had risen so terribly, when recent crops had been so good? She had little idea. Was it the price of petrol? Certainly, she knew that fertiliser had got more expensive.

It was rice planting season and the children were working alongside their parents in the paddy fields, putting in the seedlings. In normal times, a day's labour in the paddy field would pay about 3,000 riel (40p). 'A family of six eats at least two kilos of rice a day,' Han Sophan, a community leader, told me. 'Two years ago, you could buy that much rice if you worked for a day and have a little extra money. Now 3,000 riel buys only just half the rice needed to feed a normal family. So what can people do? Adults are going hungry so the children can eat.'

The sad truth was that Cambodia could have fed itself on its own rice. But economic reforms forced onto the government by the international financial institutions had closed the state-run rice storage and rice processing facilities, all in the name of free trade. 'It's true that the entire country could feed itself with its paddy rice [raw, unmilled rice]. But what it lacks is the ability to store or process that rice,' Sumie Arima, a trade policy analyst for Oxfam, told me.

Among Cambodia's problems is the fact that its rice growing techniques remain medieval, compared with the neighbouring countries. In Thailand, mechanisation, irrigation and modern growing techniques make rice paddy as much as five times as productive; in Vietnam it's normal to reap two or three crops a year, compared to Cambodia's one.

Mr Yang moaned that, despite all the investment put into the country by the World Bank and other international institutions as they attempted to modernise the country, no one thought to build up the rice-processing industry, or even increase storage capacity. 'I don't understand why we can't invest in these facilities; it makes profit for the farmers, for the country and provides jobs.'

IDEOLOGY BEFORE FOOD

The truth, of course, is that, as their ideology dictates, the expert Western economists prevented the Cambodian government from making such public investments: these things should be left to the private sector and the free market. The problem is, though, that this prescription seems to have left Cambodia in the lurch where it matters most: providing the security of adequate, affordable food for its people.

As the prices soared and tumbled on the commodities markets in Chicago and London, someone was getting rich out of Cambodian rice. But it was not Cambodia. When prices eventually stabilised the following year, people became aware that the key factor in the price rises had been the markets themselves. Those markets were not that concerned by the actual supply of rice, or the lack of it – more significant was the way the price could be manipulated.

Trade 'liberalisation' in the early 2000s dispensed with the old national and transnational rice funds, who could intervene, selling or buying when prices looked shaky. Speculators realised that this essential food crop could be moved just like pork bellies, soya oil or any other commodity. Free markets are risky places; people like Lai Phon, and millions of other consumers and producers of rice have no power in them at all.

The world was well aware that problems in rice supply could cause enormous social and political problems in the countries so dependent on the staple. Two appalling famines had occurred in the twentieth century because of political manipulation. Rice shortages in Bangladesh killed three million in a famine in 1974, shortly after the country gained its independence. Being an impoverished new state combined with the aftermath of a terrible war and floods were blamed for the rice prices rise. But, ultimately, it was not shortage of rice but poor administration of the stocks and the export of Bangladesh's rice to service its foreign debts that were seen as the cause.

Only thirty years before the same people had suffered a famine that killed at least two million because of mismanagement by a British government more interested in keeping rice out of the hands

Rice hero – Yuan Longping

In 1973, the Chinese scientist Dr Yuan Longping successfully cultivated the first high-yielding hybrid rice. He combined existing rice varieties with wild strains to produce a super-rice, now used in sixty per cent of Chinese rice production. His work was inspired by his frustration, as a young agricultural scientist, at the lack of means to help his people in the great famine that hit China in the late 1950s.

Increasing yield by thirty per cent, his hybrid rice suppresses the reproduction systems of the plants and is thought to provide food for seventy million additional people a year in China. Hybrids made through his techniques are now used throughout the world. In 2004, Dr Yuan was awarded the World Food Award, given for outstanding work for increasing the quality or quantity of food.

'If half of the rice-growing areas in the world are replaced with hybrid rice varieties with a two ton per hectare yield advantage, it is estimated that total global rice production would increase by another 150 million tons annually. This could feed 400–500 million more people each year. This would truly be a significant contribution to ensure food security and peace all over the world,' wrote Dr Yuan in 2018.

Two-to-six tons of rice per hectare is usual throughout the world today. But shortly before his death aged ninety, Yuan Longping was working with his team on a third-generation hybrid that in trials was giving 13.6 tonnes of rice per hectare.

of its Second World War enemy the Japanese, who were occupying neighbouring Burma and giving it to people directly concerned with the war effort.

What is the answer? Today, national governments are not powerful enough to manage the price of such an important global commodity and the price of rice is still dictated by market forces that have little to do with actual rice availability. While prices have been a little more stable in the last decade, there was another terrifying spike in the rice price in May 2020, driven by fears in global markets about the wider effects of coronavirus. Rice production has, in fact, grown at a steady one or two per cent overall in recent years.

SOLVING THE RICE PROBLEM

With governments seemingly unable to curb the speculators in the commodity markets, many people ask whether technological solutions are the way forward. In the late twentieth century, the so-called 'Green Revolution' in Asia saw rice yields double in many countries. The key was use of artificial fertilisers and the arrival of new commercial hybrid rice species.

Efforts along these lines continue. The International Rice Research Institute regularly trumpets new irrigation and plantation techniques and new genetic developments. As a result, overall rice production and yield per hectare is going up. Genetic modification has promised much: Golden Rice, adapted to deliver vitamin A to poor populations in need of it, is being rolled out in much of Africa – addressing the 120 million children at risk of blindness from vitamin A deficiency. GM techniques for harnessing rice plants tolerant of deep water or arid soil are being developed to counter the primary effects of climate change in rice-growing regions – like drought and flood. Critics of hi-tech solutions make the point that poverty is the key reason people cannot get the nutrition they need: why do we not address that first?

Among the many sceptics is agricultural scientist and trade ethics expert Geoff Tansey, who told *The Food Programme*, 'There's no technological magic bullet to [food insecurity around rice] – the solution to these problems is not about simple technological innovation. We need much broader ranges of innovation, in institutions and societies, in the rules that frame how trade is handled, about storage capacity, about reserve stocks.'

Tansey says that too often technological innovations only benefit 'the big labs', who are interested in commercial exploitation of the new strains that they patent and which often demand expensive proprietary fertilisers and pesticides to be effective.

RICE AND CLIMATE

As so often with meat and crops, the most productive methods of farming are the least planet-friendly. Rice is particularly problematic. The seasonal flooding system and vast quantities of fertiliser used in paddy farming results in the release of nitrous oxide and methane (the latter from bacteria in the soil stimulated by the floods). As a result, rice farming is said to account for 2.5 per cent of all climate-affecting greenhouse gases, far more than wheat or soy. Beef and dairy farming is responsible for 3.4 per cent. Growing rice in flooded paddy encourages bacteria, which produce methane, a far more climate-damaging gas than carbon dioxide.

Rice is responsible for far more greenhouse gas production than bread

Chicken tikka masala with rice

Rice (479)

Chicken tikka masala (1889)

Chickpea tikka masala with rice

Rice (479)

Chickpea tikka masala (604)

Chickpea tikka masala with bread

White bread (120)

Chickpea tikka masala (604)

Total 2368g CO_2e Total 1083g CO_2e Total 724g CO_2e

One portion of a curry supper:

Chicken tikka masala and rice 2368g CO_2e

Chickpea tikka masala and rice 1083g CO_2e

Chickpea tikka masala and naan bread 724g CO_2e

If you don't want to eat bread, then cous cous, also made from wheat, is another option. Or we could switch to 'low-impact rice', grown dry, without the problems paddy field farming entails. This also involves more careful use of fertiliser. But, once again, these choices are currently available only to the richest of us and rice, as we have seen, is above all others a food of the poorest on the planet.

Data from S.J. Bridle, *Food and Climate Change Without the Hot Air* (UIT Cambridge, 2020), based on European farming systems

Climate change is already affecting rice production – as with other cereals, some sixty per cent of the land currently used to produce it may become untenable. Desertification is affecting rice growers in West Africa, China and India. Ever more floods in Bangladesh could, under some sea-level-rise scenarios, reduce this poorest country's landmass by seventy per cent within a century.

BROWN VS WHITE

The one type of rice that chef Khamphu Pholsri never ate in his Thailand childhood is brown – the unmilled or partially milled grain. Similarly, it does not really exist as an ordinary food in China, Japan or India. 'Only the poorest people ate unmilled rice: it was a sign of being of the lowest class, a humiliation,' he says. Now, his partner, also Thai, tries to get him to eat it. 'She says it will lower my cholesterol. I agree the taste is interesting – but it's not *khao suay*: beautiful white rice.'

Attitudes to bread have gone through similar colour/class debates over the centuries. As we saw in chapter one, white bread has been the bread of the rich and now has an intermediate status: the bread of the poor or the self-indulgent. Unless, of course, you are eating a French baguette. Brown or 'whole' rice does indeed, through retaining the bran of its shell, contain more nutrients, as does semi-milled wheat flour. Concern about losing the nutrients led the United States to order 'fortification' of white rice – and white flour – with vitamins and iron during the Second World War.

Brown rice's level of minerals, including iron and manganese, is about double that of white rice, and vitamins like niacin and vitamin B6 are higher, too. But people who already consume healthy mixed diets have little to fear from lack of these. The key difference is in the amount of fibre: 100g cooked white rice contains 0.4g of digestion-aiding fibre; brown rice contains three times as much – and eating fibre probably can reduce your bad cholesterol. However, eating one apple will more than make up the deficit of a white-rice meal.

RICE: A DEMYSTIFICATION

For some reason, rice has an incomprehensible language attached to it, by growers, marketing people and cooks. Here's the beginnings of a glossary:

Arborio and other risotto rice. Arborio is named after a town in the Italian Piedmont, but is grown all over the world. It is simply the largest absorbent rice grain used for risotto. 'Superfino' on the packet does not mean it is better, simply that the grains are more than 6.4mm long: Carnaroli and Vialone Nano are shorter and preferred by many chefs.

Basmati. This means fragrant in Hindi. It is a long-grained rice grown across north India and Pakistan. Patna rice is similar.

Brown rice. This still has part of the seed's coat; the bran, germ and aleurone layers are still present and so the rice contains more minerals and other nutrients. It won't absorb water as easily as white rice and thus needs longer cooking. Part-polished rice also exists – in Italy it is called *semi-lavorato*.

Enriched or fortified. One in twenty grains has been given a soluble coat containing vitamins, usually niacin, riboflavin, iron and thiamine.

Parboiled or converted. This is an ancient technique still used in some modern rice brands, like Uncle Ben's. Since the first boiling is done before milling, more nutrients and taste may be retained and the resulting rice grains are more likely to stay separated, or 'fluffy'.

Quick cook and **boil in the bag.** The milled rice has been parboiled, breaking the cell walls so the final process can be done quickly. Likely to mean a loss of taste and texture.

Red rice. The variety of red rice from the marshy Camargue in south-west France is famous but it is grown across the world from Malaysia to Brazil as well. The red comes from the husk and because, as with brown rice, this isn't removed through the normal hulling and polishing that makes white rice, the healthy nutrients remain.

Short-grain rice. This is more common in East Asia, where the rice is prized for its tendency to clump. It is used in sushi.

Sticky or glutinous rice. This is very high in amylopectin, a starch that makes the grains cling to each other; it is usually steamed, as too much water will make it disintegrate. Its chewy texture makes it the preferred rice in northern Thailand and Vietnam, where it is also made into chewy cakes and puddings. Sticky rice cakes with mashed bean or meat fillings – easily transported and long-lasting in a leaf wrapper – are said to have won the Vietnam war for the North Vietnamese.

White rice. This is fully milled, so all the protective coat has been removed. Fine wire brushes will have polished it. White rice keeps longer and better than other types but lacks minerals, vitamins and fibre.

Wild rice. Actually seeds from the *Zizania* species of wetland grass, so this is not really rice at all, according to botanists. It is sold usually mixed into white rice to add taste and interest. Native American peoples like the Ojibwe traditionally harvest wild rice by canoe.

CHAPTER 11

THE DUKE OF DEVONSHIRE'S CURIOSITY

BANANA

'Some day, somebody will decide to take a risk and figure out how to import these better bananas … Gourmet bananas! I promise, take a bite of an 'ice cream' banana or a Fijian *fehi* banana or a Congolese *ibota ibota*, and you will pay money for that banana again.'

Banana historian Dan Koeppel,
The Food Programme, 2013

What a wonderful thing is the banana. The perfect fruit: tasty, neatly delivered in a biodegradable packet, full of potassium, vitamins, fibre and a load of healthy carbohydrate. They can play a worthy role in geoeconomics, too. The trees must be grown in the tropics; they need at least fourteen consecutive frost-free months to grow to the point of producing fruit. This enables the transfer of wealth from richer, colder nations that want bananas, to poorer, warmer ones that can grow them. What could go wrong?

BANANA DOMINATION

We love bananas, above all other fruit. The average Briton, adult and child, gets through two a week – mostly from South America and all of one type, the Cavendish. They are cheap and, perhaps as a result, we throw away a shaming 1.4 million of them a day. By weight, bananas are by far the world's biggest fruit crop (watermelons come next, followed by apples and oranges). They are eaten everywhere, usually raw in northern countries and cooked in the tropics. Plantains, famously fried in the West Indies, are simply another member of the *Musa* genus, though their sugar content is about a third that of a ripe dessert banana, the ones we eat most often in Europe.

One hundred and thirty years ago, hardly anyone in Britain had seen a banana. It was a botanical curiosity. The few recipes from the mid-nineteenth century suggest it be served sliced or chopped up: the Victorians had a big problem with its shape. As late as 1872, Jules Verne, in his novel *Around the World in Eighty Days*, had to explain what a banana was – 'as healthy as bread and succulent as cream' – to his readers. It was not until 1901 that the British pioneer banana shippers, Elders & Fyffes, started regular shipments back to Britain, in cooled freighters, from Jamaica and the Canary Islands.

In the early twentieth century, modern shipping and industrial agriculture swiftly took the banana to household staple. That stopped abruptly when the Second World War began, as the Atlantic shipping routes became too dangerous and too important for ships threatened by German submarines to be used for luxuries. In December 1945, with war rationing still in place, the British government decreed that every child deserved a peace-celebrating treat: they should get to have a banana. No one under ten years old could, presumably, remember what they tasted like. Ten million bananas arrived on the last day of the year from the West Indies, on Elders & Fyffes' ship, SS *Tilapia*.

When three bananas arrived at the Somerset home of the novelist Evelyn Waugh he summoned his three children, all under nine, and sliced up the bananas with sugar and cream. Waugh liked to tell his friends that he disliked his 'boring' offspring.

Perhaps to challenge them into being more interesting, he ate all the bananas under the children's gaze. 'It would be absurd to say that I never forgave him, but he was permanently marked down in my estimation from that moment,' his son Auberon Waugh later wrote.

BANANA EMPIRES

The trading system that brought the bananas to the Waugh family began only sixty years earlier, with the entrepreneur Andrew Preston: he is to the banana what Christopher Columbus is to the chilli or Sir Walter Raleigh to tobacco. He was working in the Boston fruit and vegetable trade when, in 1870, he met a merchant ship's captain, Lorenzo Dow Baker, newly arrived from Jamaica with a bunch of extraordinary yellow fruit like nothing Preston had seen before. He was captivated. The two men struck a deal – if Captain Baker could find room among his regular cargoes for bananas, Preston would work up customers for them in the United States.

Dan Koeppel, author of a history of the fruit, explains: 'He had this crazy idea that people would want to eat bananas. Nobody really knew what a banana was and the idea of eating them whole was taboo because of the suggestive shape. He was a marketing genius. One of the promotional things he did was to have a series of postcards, photographed, of elegantly dressed women, sitting in gardens and parlours, holding bananas in their hands and eating them and bringing them to their lips – and it really worked.'

Slowly, Americans took up the banana, while Preston and Baker learnt the complexities of bringing fresh fruit nearly 2,000 miles to market. In the 1880s, they set up the Boston Fruit Company. The strategy was to 'make bananas half the price of apples', and in doing so they built the world's first intercontinental supply chain for fresh produce.

Cheap production in Latin America and the Caribbean – 'turning the jungle into a factory' – was a key. But Preston also needed a reliable, uniform banana from among the 1,000 or so species then available. 'It had to be high yielding, sweet tasting and good looking, tough enough to travel, ripen in such a way that it had a long shelf life, and growable in large plantations,' as Dan Saladino put it on *The Food Programme*.

In the countries around the Caribbean, the new banana trade first appeared to be an enormous boon. Many of them had been in dire poverty since the collapse of the sugar industry after the end of slavery in 1833. The British government, which by 1901 had become concerned about living conditions in its Caribbean colonies, subsidised the new trade. But Preston and Baker's company soon swallowed up others and became United Fruit Company. They exploited the workforces, bought great chunks of land in Latin America and began interfering in the politics of the countries that became dependent on banana dollars. By 1924, when Preston died, the company operated a near monopoly (it had bought up the British Elders & Fyffes), employing 67,000 people and owning eighty steamships and millions of acres.

The banana corporations bribed their way into control in these countries – hence the term 'banana republic', first coined in 1904 to describe Honduras after United Fruit Company had essentially taken control of it. The violent suppression of plantation workers – notably in Colombia in 1928 – who were demanding better pay and conditions became a recurring part of the banana story. In 1954, the CIA conspired with United Fruit Company to enable a right-wing coup in Guatemala after a populist government demanded the company pay more tax.

The banana republic syndrome showed how dangerous it was for poor and fragile countries to be dependent on a single crop for export – especially when faced with rapacious corporations and their paid allies in government. Later in the century, the flood of money engendered by another South American product much desired in the richer world – cocaine – was to wreak further damage to lives and democracy in the region.

'Look at the mess we've got ourselves into, just because we invited a gringo to eat some bananas.'

Gabriel García Márquez in his historical novel of Colombia, *One Hundred Years of Solitude*

Like other poor nations whose raw materials are the subject of great demand in the rich world, little of the value generated by the great growth in banana consumption has come back to the ordinary people in the countries that grow bananas. The countries remain among the poorer, and labour conditions on banana plantations remain troubling in some of them: in 2002 Human Rights Watch found eight-year-old children working on plantations in Ecuador, the world's biggest banana exporter and source of many eaten in Britain. In 2019, certified 'fair trade' bananas amounted to less than four per cent of a global banana business of 21 million tonnes.

Even today, bananas are said to be the most profitable item in British supermarkets. Just five companies – Dole, Del Monte, Chiquita, Fyffes and Noboa – control eighty per cent of the international trade. Chiquita is the inheritor of Preston and Captain Baker's United Fruit Company.

Cook your bananas

In Britain, apart from the occasional banana bread, we tend not to cook bananas. That may be a mistake, not least because it's a way of using them when they're past their pristine greeny yellow. Banana fritters are a favourite across Asia, though they don't look like the soggy ones we know. In Thailand, close to the banana's origins, banana fritters – *kao mao tort* – are much more complex and interesting than ours. Their crunchy coating includes lime, crushed roasted rice and coconut flakes. People are more adventurous with the banana in Thai kitchens than we are here. Salt is used to bring out the underlying flavours: there's a lovely Thai dessert, *makaam guan*, which blends puréed banana with coconut, lime and a lot of salt.

MONOCROPS

The first banana the American companies espoused was the Gros Michel (a.k.a. 'Big Mike') and it was, by all accounts, a tastier and tougher banana than today's staple, the Cavendish. Its seeds had been brought to the Caribbean from South East Asia during the slavery era. But the Gros Michel was susceptible to a devastating mould disease and, in the mid-1950s, all banana production from the Caribbean and Latin America for North America and Europe shifted to a species unaffected by the mould, though it required the use of a lot of pesticide. While other tropical countries still grow many different species for their own use, that banana, the Cavendish, still dominates the world trade. It makes up nearly half of all banana production.

The Cavendish has had an extraordinary journey. Originally from China, it carries the family name of the Dukes of Devonshire, who brought it from Mauritius and grew it in their greenhouses

Think you know bananas?

In Britain, we only eat one variety but there are over one thousand: not all yellow, and not all banana flavoured. The Silk or Tundan banana, grown in West Africa and East Asia, has a tangy hint of apple. There are purple, red (with a faint raspberry flavour), orange, gold and pink bananas, and of course the prized Blue Java banana – it is the colour of a clear dawn sky and it tastes of vanilla ice cream. Some bananas have the most extraordinary aroma. There is a Chinese variety called 'You Can Smell It from the Next Mountain'.

at Chatsworth House from the 1830s. The Devonshires served it as an exotic oddity at dinner parties. A missionary took some of the Chatsworth plants to the Pacific islands of Samoa. Thence the banana made its way to Trinidad, where government botanists introduced it to the plantations. Its success there took it back to East Asia and the Pacific again.

'Every Cavendish banana,' says Dan Koeppel, 'is basically a genetic clone of every other one: this provides the uniformity that we want. Bananas are extremely reliable but because every banana is an identical twin it is susceptible to the same diseases and they hit hard, destroying banana plantations around the world.'

YES, WE HAVE NO BANANAS

A virulent fungus, related to the one that destroyed Gros Michel, emerged in the 1990s. It has now devastated the Cavendish-growing banana industry across parts of Africa, Australia and much of East Asia. A mere speck on a visitor's shoe will infect leaves and stem, turning what should be green to yellow and then black and slimy. Tropical Race 4 (TR4), as it's known (the fungus which brought down the Gros Michel was called Tropical Race 1), will quickly lay an entire plantation to waste; even plants that briefly recover remain infectious. The fact that the plants are 'monocultural' (all genetically identical) means they all will die in the same way, at the same rate. The land cannot be used again for bananas – or, at least, not for the Cavendish. TR4 wiped out almost all Malaysia's plantations in less than a decade.

By 2021, the fungus was present in eighteen countries, including India. Its arrival in Latin America, long dreaded, happened in Colombia in 2019. So far, it appears to have been contained there but the continent's banana farmers live in fear. The obvious solution, widening the choice of bananas beyond Cavendish (which is ninety-nine per cent of all the bananas exported around the world), seems to be a non-starter: the chief problem is resistance from us, the consumers. We want bananas cheap and uniform, and that is what the Cavendish gives.

The race to find a solution to TR4 started years ago and is centred on breeding resistant strains, first by looking at wild banana varieties that are not susceptible to the fungus. This is urgent – many Latin American countries are still hugely dependent on banana exports, and in other countries banana is a crucial source of nutrition. But genetic alteration through traditional breeding is slow: it may take twenty years to find the right mix.

MODIFYING THE
BANANA IN THE LAB

In Queensland, Australia, the disease has hit the banana industry hard. Research into TR4 solutions is moving faster there than in Europe because Australia is not bound by the same strict regulations on GM and gene editing. DNA from other species and other plants or bacteria are being inserted into the Cavendish genome, with the hope of designing an improved version of the banana. 'Within the next twenty years, if you're going to eat Cavendish bananas, they're probably going to be genetically modified,' Professor James Dale, who is leading the research, told *The Food Programme* in 2019. He has experimented with genes from rice, thale cress and even a type of worm.

Banana nutrition

Bananas are an amazing package of dietary goodness, a source of fibre, minerals, vitamins and various antioxidants and phytonutrients. One 100g banana will provide around ten per cent of your recommended daily intake of potassium, vitamin C, magnesium, copper and manganese, and thirty-three per cent of your intake of vitamin B6. (Though smoking banana skins won't get you high, despite the beliefs of fans of Donovan's song 'Mellow Yellow'.) That 100g banana delivers about ninety calories, depending on ripeness. A green one is about eighty per cent starch, when you exclude the water content, but during ripening almost all of that will be converted to sugar.

Dale, who is funded by the Australian government, defends the notion that GM is the only way forward: 'Bananas are probably the safest crop that you could possibly think of from a genetic modification perspective. They are essentially sterile, so there's no transgene flow [leading to contamination of other plants] through pollen.'

Backed by the Bill and Melinda Gates Foundation, Professor Dale is also using genetic modification to help African countries where vitamin A deficiency in bananas is a huge problem. In Uganda, people eat half a kilo of cooked banana a day but the most common variety, matoki, is short on vitamins. Trials of the improved banana have now begun across East Africa.

But do Africans want the Gates Foundation's GM bananas – which, for a start, will have orange flesh? Edward Mukiibi, a farmer and agronomist in Uganda, points out that he and his fellow farmers have fifty-two native banana varieties, each with a different character. They know how to use different types for different challenges. 'The transgenic banana is uncalled for,' says Mukiibi, who is also vice president of Slow Food International. 'Farmers are the first scientists, who have evolved these varieties over generations. They may not work in a laboratory with a white coat: they work with their tools and their knowledge. That's not being against science but being against bad science, against science that is dangerous for the future of food on the African continent.'

'The answer is really not just a GM banana,' says Dan Koeppel, 'but bringing variety to the supermarket. There are amazing bananas out there. The Cavendish banana is just a lousy banana. In India it's called the hotel banana. It's just served to tourists. Just as we've done with citrus, with apples, the idea would be to bring diversity to the market and insulate against any one banana being wiped out.'

A BANANA-SHAPED FUTURE

Bananas are by no means the worst offender, in climate damaging terms, among fruits we import. Shipping by refrigerated boat is far less expensive, in greenhouse gases, than the air freight that can be used, not least for strawberries in winter. Despite the fact that they come in bunches and their own handy and eminently biodegradable protective package, they are boxed and plastic-bagged for retail. The heavy cardboard boxes are, according to Professor Sarah Randle in her *Food and Climate Change*, more of an issue in terms of environmental cost than the plastic wrappers used by supermarkets. Nonetheless, though an imported banana is responsible for twice the emissions of a British-grown apple, it still amounts to only two per cent of Randle's suggested daily greenhouse gas budget.

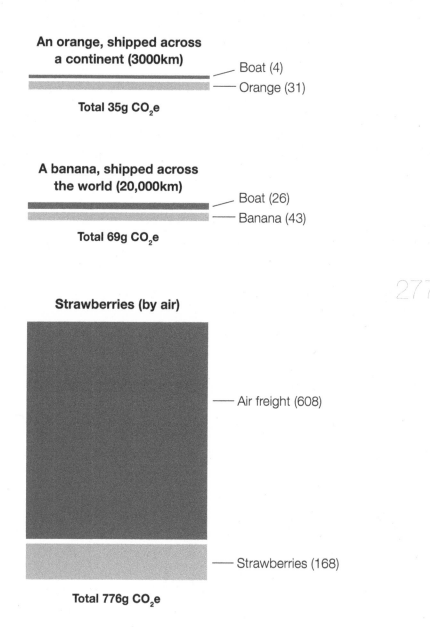

An orange, shipped across a continent (3000km)

Boat (4)
Orange (31)

Total 35g CO$_2$e

A banana, shipped across the world (20,000km)

Boat (26)
Banana (43)

Total 69g CO$_2$e

Strawberries (by air)

Air freight (608)

Strawberries (168)

Total 776g CO$_2$e

Data from S.J. Bridle's *Food and Climate Change Without the Hot Air* (UIT Cambridge, 2020)

As with so many of the foods in this book, it looks as though sorting out the problems with the banana business is about going back to basics. No one could suggest that the fruit should stay in the tropical countries where they grow: exporting them is an important income for some of the world's poorer countries. But growing just one type of banana is plainly dangerous, both to those countries and to the planet. There is no reason for doing so other than to keep prices down and profits up. Can we pay a little more and can retailers take a little less to ensure a healthy banana trade that is good for us and the people who grow them?

CHAPTER 12

THE UBIQUITOUS BEAN

SOY

'In the last five decades this small, yellow-brown bean
has changed the way we eat. It's both celebrated
and controversial. It underpins much of the
modern industrial food system and yet
it's an ancient crop highly prized in
traditional Asian food cultures.
So it's a bean of extremes and
one full of surprises ...'

The Food Programme,
2012

There is no ingredient so embedded in our food and cooking lives about which we are quite so ignorant. How many people in Europe know what a soy bean looks like or how its plant grows? Yet soy is ubiquitous: from it we get one of the most used cooking oils and the most important filler protein for everything from the feed our animals eat to the pre-prepared meals and snacks upon which we – the British in particular – are now so dependent.

English adopted the words soy and soya from *shoyu*, the salty Japanese sauce made from soy beans – that's how Westerners first encountered the product of the soy bean (which the Japanese call *daiku*). There the sauce was a by-product of the more important task of making miso, the fermented paste made from the beans that is a cornerstone of East Asian stews, soups and fry-ups.

The British philosopher John Locke first noted *shoyu* sauce in his journal in 1679. He anglicised it as 'saio'. He was writing about the fascinating novel ingredients that British East India Company ships were bringing home from their Asian voyages; Locke mentions a sauce made of mango, too. Other writers referred to 'catch-up' from the East Indies, probably a soy-based sauce. It was pricey: one guinea (£1.05) a bottle in London in the late seventeenth century, which is £174 today.

The soy bean has an ancient and noble role in Asian cuisine, in staples like tofu, tempeh and umami-laden sauces. That has nothing to do with its unique rise as a global food crop at the end of the twentieth century. From the industrial food age, the only Asian invention is instant noodles (essentially wheat flour, soy oil and flavourings), which were first made in Japan in the 1940s. Very little of the soy we eat – much of it in oils, 'meat extenders' and other filler products – is visible. Nonetheless, the average Briton consumes over a kilo of it a week.

The bean that has come to dominate industrialised agriculture across the world looks pretty boring: small, round and vaguely brown in its stable, dry state. Recently, we've got used to the fresh, immature beans, green and crunchy and called edamame, in Japanese restaurants. To a Western gardener, the plant doesn't look unlike French or runner beans: the hairy pods hang in clusters from a similar tough lime-green stem. But what happens to the soy after harvest is where it all becomes extraordinary.

SOY FOR EVERYTHING

Soy's late-twentieth-century ascent to world domination owes a lot to poverty in Brazil, to rising demand in the richest countries for ever-cheaper meat and eggs, and to hideous diseases like bovine spongiform encephalopathy (BSE).

To take the last first, BSE was the direct result from the profit-hungry process of turning farm animals cannibal: feeding the spines, brains and other unwanted parts of slaughtered creatures back to them. Even the faeces of battery chickens went for processing into animal feed. This practice developed in the 1970s as super-cows began to emerge through selective breeding. They could give three times the milk that their ancestors did, but grass was just not enough to provide the energy needed. So fish oil made from anchovies was made into cakes to supplement their normal feed. In the United States, feed-lot farms – whose 'more efficient' cattle spend their entire lives in sheds – emerged where such food made up most if not all the diet. When the anchovy population of South America collapsed, the next bright idea was to use the protein-rich waste from slaughterhouses. The feed industry kept the source of it secret: they persuaded government that sacks of feed need only list the offal as 'protein'. It was a disaster.

When in the 1980s veterinary surgeons began to discuss a connection between this novel cheap feed and the 'staggers' disease increasingly seen in cattle, journalists for *The Food Programme* were among the first to dig into the story and take it to the public. By then, in 1988, public health scientists were concerned but it took a decade for government to act decisively on the slaughterhouses, the feed and the risk to us.

BSE had jumped to cattle from sheep, where it is known as scrapie, and it was soon found other animals were at risk, including humans. The diseased tissue was found not only to turn the brains of the animals to mush but also to cross via the digestive system to our brains where, as Creutzfeldt-Jakob disease (CJD), it has the same effect on humans. Hundreds of thousands of cattle had to be slaughtered in Britain (and some across Northern Europe) in the 1990s to eradicate the disease. Furthermore, a new cheap protein source had to be found.

Brazil was the first country to step in. Having suffered a series of famines, in the 1960s the country started to look for a way to exploit the *cerrado*, or 'closed land' – the vast, arid pampas of its central plains. The state agricultural institute decided to develop a strain of soy bean plant, then a tropical species, that would tolerate these conditions. One of soy's advantages is that it is a nitrogen-fixer, like clover and alfalfa: it actually increases the amount of nitrogen in the soil where it is grown. Using it as a secondary crop could help farmers grow wheat, then Brazil's most important crop.

A vast effort began to transform millions of hectares of ancient grassland into vast mechanised farms. The huge new crop of cheap soy found ready markets among livestock farmers in China and Europe. So great was the demand that the European Union, which has stringent rules on genetically modified imports, allows in soy for animal feed. By the early 2000s, Brazil was overtaking the United States as the world's biggest soy bean exporter, closely followed by its South American neighbour, Argentina. Three quarters of the soy was – and still is – destined for cheap animal feed.

OLD-SCHOOL SOY FOOD

The companies today that are using soy for burgers and sausages aimed at the booming vegan market are not doing anything very new. In Asia, processing soy to make solid foods is a treasured and ancient technique.

There are many ways of turning the indigestible and uninteresting dried beans into pleasures. Two thousand years ago, the Chinese began puréeing and straining boiled beans to make a milk-like soup, then adding sea salts to coagulate the soy. This becomes

'The most usual, common, and cheap sort of food all China abounds in, and which all in that empire eat, from the Emperor to the meanest Chinese … is called Teu Fu, that is paste of kidney beans [Navarrete is mistaken: he means soy beans]. They drew the milk out of the kidney beans, and turning it, make great cakes of it like cheeses, as big as a large sieve, and five or six fingers thick. All the mass is white as the very snow … Alone it is insipid, but very good dressed and excellent fried in butter.'

Domingo Fernandez Navarrete, a Spanish missionary in China in the 1650s soft-textured lumps known in English as bean curd or in Chinese and Japanese as tofu: the bland white cubes in miso soup, or the scrambled-egg-like substance in a stir-fry. It remains the preferred way of eating solids from soy beans across Northern Asia.

Tofu comes in many forms. There's a huge difference between factory-made, pasteurised tofu blocks you can buy in Asian supermarkets and the delicate, silky strands of freshly made tofu, known as *tofu-hua* and often eaten as part of a dessert in East Asian countries. 'Stinky tofu', which has been matured with a bacterial culture, like a cheese, is rotten-feet pungent: it's usually eaten deep fried. In Korean towns, artisan tofu sellers tour the streets with blocks of it, ringing a bell for customers like an ice cream van.

The soy 'milk' that has become a twenty-first century Western habit has ancient origins. It was sold by street vendors in China from the eighteenth century and first marketed in the West in France and the United States after the Second World War as healthier than dairy milk. The American version was called Soy-Lac because of objections from the dairy industry to use of the word milk. It's an important ingredient in many cuisines across Asia – a dessert of iced soy milk, sweet red beans and ice cream is a Korean favourite, according to the chefs Jordan Bourke and Rejina Pyo in their best-seller, *Our Korean Kitchen*.

FERMENTING SOY

Further south in Asia, another tradition emerged using fast fermentation. In Java, people learnt to make a solid soy called tempeh – probably before tofu arrived from China, though the trajectories are hotly debated by nationalistic food historians. It is made by soaking and then boiling dry soy beans and the resulting porridge is made 'live' by adding a starter culture – just like with bread or cheese. The bacteria grows on the warm, damp mash of beans, breaking them down and fusing them together into a fluffy, chewy white mass.

Tempeh is best fried, when it tastes nutty and as close to meat as anything that hasn't been deliberately flavoured to mimic beefburger. It's the base of a host of wonderful Indonesian stir-fries and sauces. Indonesians treasure tempeh. It is a nation-defining dish; their Yorkshire pudding. But, historically, it was a food of the poor – unlike tofu further north – and to some modernisers at the end of the colonial period it was a source of shame. 'Don't be a tempeh society,' said President Sukarno, the first leader of independent Indonesia. 'You should rise above the soy bean.'

Tempeh is available in blocks in Asian supermarkets but it is possible to make it at home. It is a three-day process but all you need is soy beans, a warm spot, some bags to let them mature in and a starter culture. That culture, Rhizopus oligosporus, is available online. Since the finished tempeh uses the entire bean, unlike tofu, it has all sorts of nutritional advantages.

SOY SAUCE: MORE THAN LIGHT AND DARK

The best-known fermented soy product across the world is soy sauce, a fast deliverer of umami flavour and tangy saltiness that has been treasured for 400 years. Until recently, 'soy' meant the sauce for most Britons. The long process to make soy sauce starts with a fungal culture, one of the *Aspergillus* family that also serves brewing

A world of soy sauces

In East Asian cuisine, different soy sauces are used for different things in cooking and seasoning:

Light soy sauce or *jiang you, sang chow* in China, *shoyu* in Japan is the standard Chinese-style soy sauce, and is a rich, clear golden colour. To be used in routine cooking and recipes that call for unspecified 'soy sauce'.

Dark soy sauce like standard Kikkoman, the most popular brand, is much stronger and sweeter and a dark brown; don't use without tasting or instead of standard soy. May have added caramel.

Tamari is a stronger, traditionally made Japanese sauce, a by-product of miso paste, without any wheat in the making.

Kecap manis is an Indonesian staple, sweetened with palm sugar and flavoured with galangal, coconut, lime and garlic (depending upon which of the many treasured recipes has been used).

Saishikomi is a Japanese speciality made by fermenting the initial soy sauce again. Strong, complex, rich in flavour.

Shiro is a white soy sauce made mainly from wheat. Delicate flavour for clear soups or dipping sauces.

Tsuru Bishio is a four-year-aged soy from an artisanal producer on the Japanese island of Shodoshima, made famous by chef Samin Nosrat on her Netflix show, *Salt, Fat, Acid, Heat*. It is expensive at £40 for half a litre but worth it, according to Nosrat – a little goes such a long way.

and bread-making. As the beans (with added wheat) break down, a host of bacteria, sugars and enzymes are released. Their complex reactions produce many different flavours: several hundred different aroma molecules have been discovered in soy sauce. Some of the best soy is matured for years, like whisky, in ancient wooden barrels.

'We look for three things: colour, aroma and taste,' the food writer Kimiko Barber revealed on a 2007 episode of *The Food Programme* on maturing and fermentation. The colour is particularly important: 'People think soy sauce should be brown or black but the chic nickname for soy sauce in Japanese is *murasaki*, which means purple. It should have a tinge of red, a tinge of amber. In cheap soy sauce the colour is often added.'

The most important effect is the production of glutamic acid – the umami factor – which also produces salts, including the flavour-enhancer monosodium glutamate (MSG) – ubiquitous in Asia, but feared and reviled in other cultures.

In the West, we are now beginning to appreciate that there is more than one soy sauce, differing in strength and tastes. Asian cooks know this very well already and select carefully which they use in their cooking. But on our supermarket shelves there are fast-track, cheap soy sauces, which do not use the slow, traditional fermentation process. These mix de-oiled, crushed soy meal with hydrolysed protein from other vegetables, treating them at high temperature under pressure with hydrochloric acid. Look out for – and avoid – bottles that list additives like corn syrup, caramel and salt. Better brands may say 'naturally brewed'. Some soy sauce is legitimately made by fermenting wheat as well, meaning it will contain gluten.

'The standard has improved beyond measure since I first came to this country in the early 1970s,' says Kimiko Barber.

'Then you'd just see one bottle of Kikkoman on the supermarket shelf.' This most popular brand is a favourite in Japan and it is now made, for Europe, in a factory in the Netherlands but it does promise that the pungent Kikkoman soy sauce is 'naturally brewed' over several months.

TWENTY-FIRST CENTURY SOY

Unprocessed soy beans are not much valued in Asia, their original home, except as emergency food in times of famine. The mature beans are hard to digest, even after soaking and boiling. Most people would only turn to them when there was no rice. It is in the factory that the value of the processed soy bean today was discovered – its principal virtues being tastelessness, high protein levels and its cheapness.

289

Very little of the soy we eat today is apparent. Here are some of the things in which it plays a role, as an oil or a solid:

- Ice cream (in the stabiliser, lecithin)
- Deep-frying in restaurants
- High-protein feed for salmon, pigs, chickens and cows
- Breakfast cereal
- Cakes, biscuits and muffins
- Bread and other baked foods
- Margarine
- Instant noodles
- Stabilisers in food processing
- 'Extenders' to bulk out cheap meat products
- Vegan 'meatalike' foods

- Vegan 'milk' and baby formula
- Vegetable oil
- Polish, like the waxes on the skin of fresh fruit
- Mayonnaise and other sauces
- Shells and the coating of medications
- (And, less edible: soap, candles, glue, industrial lubricants, bio-diesel and insulation foam)

Like the fruit of olive trees – also inedible before processing – soy beans are fabulously rich in fats, with twenty times as much as broad beans or fava beans. You can get up to 190g oil from 1kg olives or soy beans. It's not the same oil, nutritionally: one disadvantage of soy oil that science is only beginning to understand is its high concentration of omega-6 fatty acids, a problem we look at in chapter three, 'Fats of the Land'. Putting soy beans under mechanical pressure, as with olives, extracts some of the oil. But treating them with heat and petroleum-based solvents, a modern American invention, gets much more: a process not permitted in genuine olive oil, though it is used on the 'pomace', the debris remaining after olive pressing. The process leaves chemical residues in the oil.

The big difference from olives, though, is the soy bean's protein. For *The Food Programme*, Dan Saladino visited the 'hill of powdery sand' that is the flakes of newly milled soy beans at Cargill's huge plant in Liverpool's docks: every month, a ship carrying 60,000 tons of the beans arrives from South America. At the Seaforth plant, Cargill operates the UK's only crusher for soy beans. At the end of the mashing and heating treatment, there is soy meal, soy oil and lecithin (an emulsifier used in food processing and medicines) and tocopherol (vitamin E) ready to sell to the food industries – animal

290

and human. At Seaforth, Cargill processes 750,000 tons of beans a year: two-and-a-half square miles of crop every day.

When soy bean oil was first industrially extracted in the US in the 1930s, it was used in industrial paints and for heating. But it was clearly more valuable as food for humans and animals. Raising the flakes of crushed beans and then flushing them with hexane, derived from crude oil, frees the oil ready for refining at high pressure. 'At the end, it's almost clear, a slight tint, much like you'd see in the grocery store: an ingredient for mayonnaise, mustard, salad dressing or for cooking,' the Cargill technician explained.

TEN YEARS ON

When Dan Saladino visited the Cargill plant in 2012, he reported that the global crop, then 240 million tonnes, had doubled in just over fifteen years. Argentina was producing forty million tonnes a year, a crop that had doubled in size since 2002. And the extraordinary rise of soy has not slowed in the decade since: in 2020/2021, world production was up by another fifty per cent to 360 million tonnes – two thirds of it produced by the United States and Brazil. Argentina's production also grew, peaking in 2019 at 58 million tonnes, up forty-five per cent in seven years. During the last decade, on several occasions, Brazil's entire production could be accounted for by China's consumption – that's where most of the beans were shipped.

This shift was accelerated by Donald Trump's decision, for political reasons, to limit trade with China when he was president of the United States. The trade war held back the American producers, put up the world soy price and had the unintended consequence of accelerating Brazil's production of soy for export. That, in turn, meant more clearing of land for soy. Demand continued to boom. In mid-2021, world soy bean prices were back at the record highs of ten years earlier, despite the massive increase in production.

It is Brazil's destruction of the Amazon rainforest for logging that hits the headlines: a globally recognised ecological disaster, sadly accelerated under the presidency of Jair Bolsonaro. But far more forest and pastureland has been cleared for soy. In 2020, Brazil's soy bean exports earned it $28 billion – perhaps ten times what it received from exports of timber. Thirty-eight million hectares, an area bigger than Britain and Ireland combined, is currently planted with soy, mostly to serve Europe and China.

THE PROBLEM OR THE SOLUTION?

Humanity is now scarily dependent on soy, nearly eighty per cent of which we feed to our animals. The rest goes straight to our food. The task of re-engineering the global food supply system to work without the bean gets more daunting every year, as the number of people dependent on soy, as food for themselves or the animals they want to eat, grows along with the number of people on the planet.

What is the eco-cost of the great soy explosion? It is hard to measure. Farming on land already used for agriculture has negligible costs in greenhouse gases, though the drain on water resources in the dry plains of Brazil and Argentina is a huge concern. Shipping the soy beans around the world may be more damaging in terms of greenhouse gas emissions. But if the growth of soy use continues – and there's no reason to believe in 2022 that it has peaked, any more than world population has – more and more virgin land will be brought under the plough. The initial clearing of land and forest releases enormous quantities of stored carbon into the atmosphere, carbon that cannot be recaptured until the forest grows again.

Activists exposed the damage being done to the Amazon basin by ever-expanding soy production in the early 2000s. That had welcome repercussions: after pressure from consumers, an indefinite block on new farms was put in place for the Amazon basin and even as a right-wing presidency loosened controls after 2019, the moratorium appeared to be holding.

However, the Amazon deal simply shifted the corporations' interest further south, to Brazil's unique Cerrado savannah and forest, home to some five per cent of the world's species. Even before the Bolsonaro regime took power, eighty per cent of the Cerrado was unprotected. Another area of serious concern is the dry river plains of the Gran Chaco in Argentina and Paraguay – also being ploughed up for soy. 'They can't be replaced,' Jonathan Gorman of the industry-backed UK Sustainable Soy Initiative told the BBC. 'These are complicated eco-systems that are thousands of years old.'

More land seems certain to be turned over for soy as world meat consumption rises. There's no sign of that slowing, especially in Africa, India and other parts of the world where people are gradually becoming richer. In almost all societies, the more you earn, the more meat you eat.

The irony of all this is that soy could be a part of the solution to the problems of greenhouse gas emissions and climate change. If we ate even some of the soy that we feed to animals, instead of eating them, the whole narrative would change. We would need much less soy, to begin with.

Greenhouse gas emissions per kilogram of food

An 8oz (225g) beef steak contributes about 10kg in greenhouse gas emissions: one such meal once a week for a year is equivalent to four months of the annual emissions from heating and powering the average UK home. A 200g piece of soya contributes about 570g in GHGs.* These figures vary according to factors including provenance and farming methods. An 8oz steak, farmed according to European methods, contributes around 10kg of greenhouse gases, dependent on how the beef animal is raised.* Tofu contributes around fifteen times less, though if the soya for tofu is produced on newly cleared former rainforest, the cost might be around twice as high.

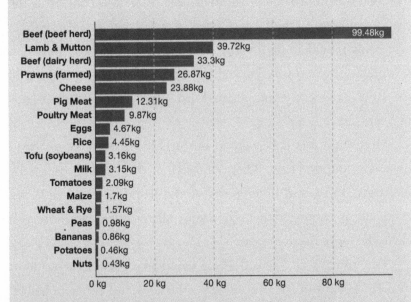

Food	GHG emissions
Beef (beef herd)	99.48kg
Lamb & Mutton	39.72kg
Beef (dairy herd)	33.3kg
Prawns (farmed)	26.87kg
Cheese	23.88kg
Pig Meat	12.31kg
Poultry Meat	9.87kg
Eggs	4.67kg
Rice	4.45kg
Tofu (soybeans)	3.16kg
Milk	3.15kg
Tomatoes	2.09kg
Maize	1.7kg
Wheat & Rye	1.57kg
Peas	0.98kg
Bananas	0.86kg
Potatoes	0.46kg
Nuts	0.43kg

Table by permission and data from J. Poore and T. Nemecek, 'Reducing food's environmental impacts through producers and consumers' (2018)

* S.J. Bridle, *Food and Climate Change without the Hot Air* (UIT Cambridge, 2020)

As the analysts at the European Institute of Innovation and Technology put it: 'Producing the same amount of protein from chicken as from soy requires three times the area of land, pork nine times and beef thirty-two times. In fact, in regards to protein intake, if the world were to swap meat protein for soy protein, agricultural deforestation would decline by as much as ninety-four per cent.' (A straight swap of soya for beef would mean missing out on other nutrients that come in animal-derived protein, of course.) It is often pointed out that the great global rise in demand for meat in recent decades has chiefly been driven by demand from China and India: as populations become richer, meat consumption inevitably rises. But in Europe we are only able to supply thirty per cent of the continent's protein needs from our own land and sea.

TURNING THE SHIP ROUND

There have been many attempts to persuade the multinational corporations that buy and process most of the world's soy to commit to basic standards and a certification scheme for stopping deforestation and promoting sustainable farming. They have not proved very effective. In 2012, a Cargill official told Dan Saladino that the company 'recognises we've a role to play' in sustainability. It was committed to working with farmers 'to improve agricultural practices'. Cargill had already joined the Round Table on Responsible Soy Association, the only global body seeking to regulate the business.

The privately owned Cargill was, and remains, one of the two biggest soy trading and processing corporations in the world. It is also one of the world's three biggest meat processors, responsible for producing a quarter of all British chicken. At its vast Avara facility outside Abergavenny, Cargill feeds soy straight from its Liverpool plant to chickens destined for McDonald's, Nando's, Tesco and Asda.

Multinational corporations make pledges easily; they find it harder to follow them through. Cargill is an economic and political force more powerful than many countries but, as a private corporation, inscrutable and hard to police. Does it keep its promises? In 2012, the company responded to international pressures by promising to stop all deforestation for soy crops by 2020. By 2019 that pledge was clearly being missed.

The food conglomerate Nestlé, also feeling the heat over the climate crisis, stopped buying Brazilian soy from Cargill in 2020 because of worries about sustainability. Addressing deforestation and difficulties around solutions, Cargill's chief sustainability officer expressed worries over stopping production in places like Brazil: 'We believe that it simply will move the problem by pushing farmers, who are interested in improving their livelihoods, to other buyers and that practices will likely continue.' By 2022, Cargill was still proclaiming its 'no deforestation' pledge but the date to achieve it had shifted to 2030.

ALCHEMY WITH SOY

The United States' most popular vegan burger comes from Impossible Foods. It is sold with the slogan 'For the health of the people and the planet' and carries one warning: 'contains soy'. That is an understatement. The ingredients list of this most successful of the vegan burgers on sale in the United States contains at least four

Impossible Burger ingredients

Water, Soy Protein Concentrate, Coconut Oil, Sunflower Oil,
Natural Flavors, Potato Protein, Methylcellulose, Yeast Extract,
Cultured Dextrose, Food Starch Modified, Soy Leghemoglobin,
Salt, Mixed Tocopherols (Antioxidant), Soy Protein Isolate, Zinc
Gluconate, Thiamine Hydrochloride (Vitamin B1), Niacin, Pyridoxine
Hydrochloride (Vitamin B6), Riboflavin (Vitamin B2), Vitamin B12.

Source: Impossible Burger

products from soy beans. The purpose of some of the soy-based ingredients on the long list is obvious: 'soy protein concentrate' will make the bulk of the pseudo-meat of the burger. Others less so. 'Leghemoglobin' is a part of the sap of plants; Impossible have patented a process using it to get what they call 'heme' to replicate meat haemoglobin, part of animal blood, and makes it look as though the burger is actually oozing blood.

Speaking to *The Food Programme* in 2018, food scientist Harold McGee told Dan Saladino, 'The first idea at Impossible was to go to soy bean fields, harvest the root systems and use this hitherto unused part of the plant to make meat flavours.' Impossible's 'meat' is expected to be available soon in Britain, if hangover legislation from our membership of the European Union over genetically modified food (which accounts for most soy grown in the States) is discarded.

Methylcellulose, a key Impossible ingredient, is made from the husks of soy or corn and is a commonly used binding agent in manufactured foods. It is made by treating the plant material with caustic salts at high temperature. It is not digestible and it's also used in cosmetics, as medicine for constipation and as a super-strong glue in electronics assembly and the fortifying of concrete. The human body appears to handle it without harm but in 2020, consumer campaigners in the United States issued an advertisement warning about its use in pseudo-meat. Impossible Foods retaliated, saying that the warning was funded by 'Big Beef' and put out a parody ad saying, 'There's poop in the ground beef we make from cows.'

'It's hardly a list of ingredients your granny would have recognised, let alone trusted,' comments Carolyn Steel in her book *Sitopia*. But what matters more, perhaps, is whether the promise is true. Is the soy burger better for your health? Is it better for the planet?

Meatalike products made from plants and on sale in supermarkets today have two to three times the saturated fat of a standard supermarket lean beefburger (about forty per cent of the official recommended daily allowance) and about half the salt content. There is roughly 3g of fibre, compared with just a scrap in the beefburger. Beyond Meat, Impossible Foods' closest rivals in the US, sell their vegan burger and other products here already – McDonald's is one of its customers. Beyond got ahead of Impossible by using material from pea plants for its base protein. (Its burger 'bleeds' on the plate, too, courtesy of beetroot juice.)

The question for many of us must be whether so many highly processed ingredients fit with our notions of healthy eating. As for the planet's health, it all comes down to the source of the vegetable protein. Soy does not have a good reputation on that with consumers, and Beyond has capitalised on that. Nonetheless, it would be very hard to follow a vegan diet today without ever eating it.

Impossible Foods' marketing makes some table-thumping points on soy. It reckons that the act of eating its 'beef' and sausage products fights back against climate change. It claims that producing an Impossible burger uses eighty-six per cent less land, eighty-seven per cent less water and produces eighty-nine per cent fewer greenhouse gases, compared with a meat burger.

'No one is cutting down rainforest so we can make tofu or plant-based burgers,' says Impossible Foods' head of impact, Rebekah Moses. The soy used is grown in the midwest of the United States and the UK's Food Standards Authority were considering an application for approval to sell the burger here as this book went to press (the EU has already turned Impossible down).

In September 2021, McDonald's in Britain followed Burger King's venture into vegeburgers and rolled out its McPlant burger, developed for it by Beyond Meat. It is made of pea and rice protein, potato starch and it also uses methylcellulose (as a stabiliser) and beetroot (for colour). *The Times*'s food critic Nick Curtis said, 'It tastes like a proper, meaty burger, with full-on umami flavour and

---

Giving up meat

The British are eating seventeen per cent less meat than a decade ago, reported *The Lancet* in October 2021 – the equivalent, for the average meat eater, of foregoing two and a half pork sausages a week. (While red meat consumption is down, we are eating slightly more chicken.) But to reach climate change targets we need to cut out more meat – the government's National Food Strategy wants us to come down thirty per cent more.

more texture and bite than most vegan patties.' Of Burger King's offering he said, 'Smarmy texture and virtually no flavour apart from ketchup: I scraped off all the toppings to check, and there was really nothing there.' Both the Burger King and McDonald's burgers were priced at only thirty pence above the meat equivalent – far cheaper than the same in supermarkets.

A SOY-RICH, MEAT-FREE FUTURE?

A few years ago, analysts were forecasting that by 2030, thirty per cent of prepared 'meat' products will be from plant sources or cultured in tanks using non-animal protein – a technique being dubbed 'molecular agriculture'. With the economic turmoil brought by the coronavirus pandemic, that now seems less achievable. Nevertheless, meat made without a living animal's involvement has seen billions of dollars of investment, with backing from the tech fortunes behind the likes of Google and Tesla. Their hope, of course, is to patent a product that the world will come to depend on. But no company is close to producing an economically viable piece of new-meat.

More promising, if that's the sort of thing you hanker for, are insect-based protein foods: in late 2021, ten companies in Britain were ready to produce crickets, mealworms and buffalo worms for animal or, eventually, human food. They need regulators to catch up with them. The insects can be fed on food and agricultural waste – the stalks of soy plants, perhaps. Fried grasshopper, pound for pound, is about three times as nutritious as beef.

For now, plant-based 'meatalike' is the only post-meat product in the shops. It is expensive: a two-pack of lean beefburgers at Tesco in autumn 2021 cost £1.50; a two-pack of pea-based Beyond Burgers of the same weight cost £4.99. All the same, both Impossible Foods and Beyond Meat have moved from speculative start-ups to maturity. The two American companies are now more than ten years old. They both raised hundreds of millions in new investment during the pandemic in 2020. Beyond Meat is in day-to-day profit, to the tune of $50 million a quarter, and its production costs have fallen twenty-five per cent. The company is trialling plant-based pepperoni with Pizza Hut, it sells plant-chicken at KFC and has a presence in eighty-eight countries worldwide, including China. Beyond Meat has set 2024 as a target date for producing a meatalike product at the same price as a real meat equivalent.

Nearly ten million Americans were on plant-based diets by 2020, compared with just 290,000 in 2004.* In the United States, the future of such companies depends not just on growing their customer base, but also on winning a battle against traditional animal agriculture in the country. The fight will be on the grand scale.

Politicians are nervous about the countryside. In the developed world, farming's lobbying muscle has meant enormous subsidies for industrial agriculture. These have kept US meat cheaper, pound for pound, than in any other rich country. (Which, of course, leaves it dependent on soy as feed for pigs, cattle and chickens.) But the need to achieve national emissions targets as the climate crisis deepens may trump industrial agriculture's lobbyists – big meat's easy ride may soon be over and big soy's reign just beginning.

---

* Survey by Ipsos Retail Performance, 'Vegan trends in the US' (2020)

CHAPTER 13

SUCH SWEET SIN

COCOA

'One thing I've always noticed, when we ask questions
and when I look at what's on the shelves in the stores,
is that the product I ate as a child is my favourite.
That says to me that nostalgia is stronger than
novelty ... Chocolate spells security.'

Marcia Mogelonsky, food trends analyst,
The Food Programme, 2021

ocoa, mixed with sugar, spice and some-times milk, makes the only world-shaping food in this book that we consume entirely for pleasure. It has some nutritional value: in its pure form, cocoa powder is rich in antioxidants and other elements that may be good for your brain and heart health. But its place here is earned by that extraordinary, obsessive human drive for happiness through the consumption of food. It has led us to some strange places and to some disastrous, damaging practices. It is a philosophical and physiological conundrum: when we eat chocolate, what are we actually seeking? And what – beyond the glorious endorphin and dopamine hit – do we get?

Like coffee, cocoa carries with it one of the extraordinary stories of human inventiveness. How on earth did we get from the unpromising beans grown on bushes in far-away countries to today's cappuccino or Easter egg? Unlike the coffee bean, which is only eaten by a few animals in the wild, the tough seed pod of the cacao tree was useful to the early peoples of equatorial America – a transportable source of energy and moisture. But there's nothing in the milky flesh that tells you it can become the base of the world's most delightful sweet snack.

Central Americans brought the beans and their tree north, through the Caribbean islands and up the isthmus to what is now Mexico. There the Aztecs were roasting and grinding cocoa beans by the sixteenth century. They made a drink for rituals and festivities, perhaps mixed with human blood. But not yet – as far as we know – sweetened. When the Spanish *conquistadores* encountered it, in great foaming jars, in 1519 at the table of the Aztec emperor Montezuma, they were told its purpose was aphrodisiac. The Spanish adventurers liked it but not everyone agreed. An Italian visitor, Girolamo Benzoni, later in the century described it as 'more suited for pigs than men'. He is the first European to describe in detail how the drink was made, by drying and roasting the cacao kernels, pounding them and mixing the paste with water and spice.

'The flavour is somewhat bitter, but it satisfies and refreshes the body without intoxicating: the Indians esteem it above everything …' He misses a crucial step, though: as with coffee and black tea, the raw ingredient needs to ferment for a few days before it is dried in the sun, toasted, peeled and pounded. The resulting chemical changes bring out secondary compounds and turn the cocoa bean into something much more interesting than the original bean.

The fermenting, foamy drinks made from the roasted beans in Central American cultures are seen as a repository of life force. In Oaxaca, southern Mexico, a woman who is pregnant is sometimes called to whip the drink, as Dan Saladino found while researching the origins of cocoa for *The Food Programme*. It is thought she has more vital energy to make the liquid foam.

305

These roasted cacao bean drinks, known as *chocolate atole*, are 'recognisably very chocolate bar', says Marcos Patchett, author of *The Secret Life of Chocolate*. 'But way more intense, more sour flavours, more robust, sometimes a little bit akin to wine, quite reminiscent of coffee, more complex in flavour and effect. For me, eating chocolate now is like the methadone version of drinking chocolate.'

Saladino and Patchett list the psychoactive properties of this chocolate: 'Caffeine, theobromine, xanthene alkaloids, which stimulate the central nervous system, as coffee and tea do. There are trace polyphenols.' The latter are responsible for the raised dopamine and serotonin levels in the brain after eating chocolate. There really is a chocolate high.

'I think this plant became so cultivated because it is so consistently and reliably pleasure enhancing,' says Patchett. '[Central America] is one of the most biodiverse regions in the world, there are all sorts of other psychoactives there, but cacao remained a really important secular and sacramental drink.' It was also about sex: 'There's lots of links to fertility goddesses and figurines with cacao pod breasts and cacao pod genitals.'

Chocolate's much shorter history in Western culture has seen it develop further complex associations and cultural rituals. From early on, its essential otherworldliness was recognised: the botanist Linnaeus named the cacao tree *Theobroma* (food of the Gods) *cacao* in his 1753 listing of plant species and that stuck. Chocolate today is about love, pleasure and giving, and it still has an ancient association with sex and sin. We have attached it, without much thinking, to Christian religious rituals: what's Easter without chocolate? It's also, not least because of relentless marketing that plays on our ambivalence, a 'guilty pleasure'. There's more to that than the average chocolate bar muncher cares to know.

CHOCOLATE AND
THE MILITARY

Unsurprisingly, the conquistadores took the bean and the techniques back across the Atlantic. They caught on at home in Spain, and preparing drinking chocolate soon became a complex, sophisticated cuisine. A 1631 recipe requires the ground cocoa beans to be mixed with chilli or pepper, anis, three different flower-spices, aniseed, cinnamon, almonds, hazelnuts and two dyes, annatto and logwood, as well as sugar.

By the late seventeenth century cocoa was known and appreciated in France, Italy and Britain, where the drink was spiced with vanilla and sugar. It was in London's coffee houses that the idea of mixing the powder with milk appears to have started. A well-travelled English physician, Henry Stubbe, wrote in 1662 of another effect of the cacao nut that he had seen in Spain, where it was taken as a 'confect', a solid: 'Being eaten at night, makes Men to wake all night-long: and is therefore good for Souldiers, that are upon the Guard.' Chocolate's caffeine content had been spotted again: Mayan warriors had discovered the value of chewing a dried cacao bean centuries earlier.

Chocolate and military enterprises continued a long association: a bar has been in a squaddie's ration pack in the British Army since the nineteenth century. In the Spanish Civil War, Republican troops were issued chocolate in sugar coatings to guard against melting. Forrest Mars, whose name is still on chocolate bars, took the idea to the Hershey Company in the United States. In 1941 Hershey brought out M&Ms, exclusively for the use of US troops.

Britain's most famous chocolate maker, Cadbury, let the public know how important its products were to the war effort. Every lifeboat for the Atlantic convoys carried its chocolate. During the Second World War, one of Cadbury's posters carried a testimonial from a soldier who made it through 'a grim artillery battle in Tunisia'. 'Lucky I had a water bottle and a bar of chocolate which kept me going those four days,' he reported. The bar would have been Dairy Milk, one of the most successful chocolate products ever devised. It was a winner for Cadbury from its launch in 1905, not least because of the greater profit margin made possible by using more milk, more sugar and less actual cocoa. The company made no secret of this. Dairy Milk was sold on the slogan 'What's made with a glass and a half of fresh milk?' Based on the same chocolate recipe, Fruit & Nut and Whole Nut followed in the 1920s: Dairy Milk has now been Britain's best seller for over 110 years.

But the recipe made difficulties for Cadbury when Britain joined what was then the European Economic Community in 1973. European regulations demanded higher volumes of cocoa to call a bar chocolate: Cadbury had found that it could substitute vegetable fats for cocoa butter and Dairy Milk was by that point less than twenty per cent cocoa solids (today it promises at least twenty-six per cent, where luxury brands of milk chocolate will have thirty-two per cent or more). It was not until 2003 that all EU countries finally permitted the 'chocolate substitute' designation to be lifted, with an agreement that Dairy Milk would be labelled 'family milk chocolate'.

SWITZERLAND AND
MODERN CHOCOLATE

In the eighteenth century, recipes for solid forms of cocoa powder began to appear. *Chocolat* – a word borrowed from Central America – was sold in France as a cake: half cocoa, half sugar, spiced with vanilla and cinnamon. The food historian Harold McGee writes that busy Frenchmen made this, with a cup of water, the first instant breakfast. Organised manufacture of blocks of treated cocoa powder came next.

Wandering Italian salesmen appeared at fairs in Switzerland with sausage-shaped bars of cocoa powder mixed with sugar and vanilla, and Swiss businessmen took notice. By the 1820s, Philippe Suchard had a chocolate factory at Serrières, powered by a water-wheel. The Swiss, despite having no cocoa-producing colonies, gained a reputation as Europe's best chocolatiers, followed closely by the Belgians.

It was in Holland that one of the most crucial steps in modern chocolate-making was made. The van Houten family, who were chocolate businessmen, pressed the beans mechanically to separate out the oils, what we know as cocoa butter. This made a less oily drinking chocolate but also, when mixed with cocoa powder and sugar, a better solid chocolate. In England, this was the basis of the 'eating chocolate' sold by J.S. Fry & Sons of Bristol. Its Chocolate Cream became a bestseller in the 1850s, and in 1873 Fry's came up with the chocolate Easter egg, drawing on the tradition of decorating hen's eggs for the feast day.

As Europe began to indulge a chocolate craze that has never ended, the Italians were going further than anyone else. In the nineteenth century, recipes appeared for liver with chocolate, aubergines with chocolate, polenta with chocolate and – according to McGee – even lasagne with a sauce of almond, walnut, anchovy and chocolate. The only echo of this that seems to have survived in British cooking is in jugged hare, the traditional stew that uses hare's blood and, in some modern iterations, chocolate too. Most people find it too rich. But Niki Segnit, in her *The Flavour Thesaurus*, reckons dark chocolate pairs well with blood. She mentions

Chocolate's greatest days

Roald Dahl, author of *Charlie and the Chocolate Factory* (1964), told how his own love of chocolate bars began. His boarding school in Derbyshire was chosen by Cadbury as a test lab: the company's research kitchen sent out samples of prototype products for the boys to try and report back on. Interviewed on a BBC TV chat show in 1989, Dahl enthused about a golden age of British chocolate: 'The cleverest was the very first one that was invented, apart from the plain bars: the Dairy Milk Flake, in 1921. Obviously someone said "Push the stuff through a thin slit and let it roll on top of itself." I think it's wonderful, it's been a bestseller for sixty years. Mars Bars are probably the greatest invention. 1932. Everyone should know these dates! Who wants to know when the Kings of England were born? It's when the chocolate was invented. Kit Kat, 1935. Maltesers, 1936. Every one of them, the great chocolate bars you eat today, were invented in the thirties … It's like the Italian Renaissance is to painting, the thirties is to chocolate.'

the Italian treat *sanguinaccio*, a soft pudding made of pig's blood, chocolate, pine nuts, raisins and sugar.

MORAL CHOCOLATE

Chocolate, Voltaire explained in his hugely influential 1759 novel *Candide*, was – along with cochineal – one of the things Christopher Columbus brought back from America. It made up, he thought, for the negative effects of another Columbus import, syphilis. Voltaire did not have the details quite right. It was not Columbus but the later Spanish adventurers who returned with cocoa and cochineal – a red dye made from a Mexican beetle. (Medical archaeo-historians are still debating the origin of syphilis.) But Voltaire was taking a position in a debate about chocolate, sex and morality that was to have surprising outcomes.

Once intellectual Europe had acknowledged its interest in chocolate and lust – an association attached to many arrivals from the 'new world' across the Atlantic, including tomatoes and potatoes – another elitist, moralising force took the new luxury on: the temperance (anti-alcohol) movement. Drinking chocolate, it was thought, could be enlisted in the crusade against the lower classes' consumption of alcohol. Until the arrival of tea, urban people did not have a happy choice when it came to refreshment: water and milk could be contaminated, so the safest everyday drink was beer or spirits, which came with their own obvious side effects. Tea, coffee and cocoa changed that.

Three Quaker families – the Frys, the Rowntrees and the Cadburys – dominated the British chocolate bar trade, all of them driven in part by their religious belief to use manufacturing

Chocolate for health

Early European chocolate drinkers were convinced it had medicinal benefits beyond the positive effect it had on libido and sexual performance. Today, cocoa's polyphenols, anti-inflammatory antioxidants, are thought to benefit the blood and circulatory systems. Research into the cocoa-drinking peoples of Central America have found their blood pressure to be much lower than others in the region who don't drink it.

business, a traditional Quaker talent, for the greater good. They found natural allies in the temperance societies that emerged in the nineteenth century. Fry's launched a British Workman's Cocoa in 1874, while the London chocolate company F. Allen & Sons printed lurid pictorial adverts showing a happy, sober, cocoa-drinking family beside a squalid room of miserable alcoholics.

Robert Lockhart, an Edinburgh businessman, a Baptist and teetotaller, opened a chain of alcohol-free 'British Workman's Public Houses' in north England and Scotland in the 1870s, later changing the name to Lockhart's Cocoa Rooms. These offered tea, coffee, aerated drinks and cocoa – and the chance to sign a no-alcohol pledge book. They were a feature of city streets until the early twentieth century. London had sixty branches of Lockhart's, including one with a temperance meeting room attached, to serve cocoa to the notoriously hard-drinking butchers of Smithfield Market.

The Cadbury family went further. They, like the other Quaker industrialists, were excessively involved in the moral lives of their workers, not just their livers. Near the factory outside Birmingham,

George Cadbury, son of the founder, John, bought land to set up a 'model village' named Bournville – still the name of one of Cadbury's chocolate bars. The idea was to 'alleviate the evils of modern, more cramped living conditions' and keep a supply of beholden workers nearby. The houses came with large gardens and health facilities, sports grounds and eventually a concert hall.

A 1903 account of the factory tells, approvingly, of how workers were encouraged to live onsite, with a dormitory set up for single women. All women staff – 2,400 of them at the time – wore white and entered the factory by a separate entrance from the men. Younger women attended gymnastics classes twice a week. Workers' health and diets were 'carefully studied' and fruit was provided in the canteen.

The anonymous writer proclaims that in every department of the huge factory 'faces are bright' and 'order and cheerfulness reign supreme'. Jobs at Cadbury were much in demand, not least because Cadbury did not pay 'piece-work', by work completed, but, as the 1903 author says, 'a satisfactory living wage to the average worker'. By 1928 there were 350 'Sunshine Houses' built by Cadbury at Bournville, rented at below market rates. Some are still run by a charitable trust; the sale of alcohol is still banned in the village.

Paternalistic and intrusive though the Cadbury regime now looks, it may have been better for the ordinary worker than Cadbury's management has proved in more recent times. The last Cadbury family member retired from the chairmanship in 1989 and the company was bought by the global food corporation Kraft in a much-disputed deal in 2010. Despite promises to conserve jobs, within a month Kraft had closed one of the Cadbury factories. The years since have seen a succession of strikes and disputes, and part of production moved to Europe, where Kraft's chocolate wing, Mondelēz, makes Toblerone and other brands. Kraft pulled out of Cadbury's deal with the Fairtrade Foundation in 2016.

SLAVE COCOA

For all their philanthropy in Birmingham, William Cadbury – grandson of the founder – and the family were not so scrupulous about the conditions of the workers at the other end of the production line. The majority of the company's cocoa came from São Tomé and Principe, a Portuguese island colony off West Africa. In 1901, William Cadbury was shown an advertisement for the sale of a plantation there, which included the workers as part of the property. This cannot have been a surprise: anti-slavery organisations had already raised alarms about Portuguese West Africa. Then, in 1905, the campaigning journalist Henry Nevinson revealed that up to 40,000 enslaved people, many of them children, were working on cocoa farms in São Tomé and Principe.

Cadbury's vacillations over their slave-grown cocoa a century ago mirror the often slippery corporate public stances around labour conditions and cocoa today. William Cadbury did visit West Africa and sought assurances from the Portuguese colonial officials: 'I should be sorry needlessly to injure a cultivation that as far as I can judge provides labour of the very best kind to be found in the tropics: at the same time we should all like to clear our hands of any responsibility for slave traffic in any form,' he said.

But it was not until 1909 that Cadbury cut ties with the colony, by which time it had found a new supply from the British Gold Coast, now Ghana. Lizzie Collingham, in her book *The Hungry*

Empire, writes that the American chocolate companies stepped in and took up the São Tomé contracts.

Cadbury's new source was plantations in the British colony of Gold Coast. Named Ghana at independence, it and neighbouring Côte d'Ivoire remain the source of sixty per cent of the world's cocoa. The spectre of slavery, and particularly forced child labour, has lingered over these and other West African cocoa plantations ever since: during the Second World War, cocoa farmers in British colonies were pushed to increase production as a patriotic duty (the cocoa was shipped to the United States to counter Britain's war debts). Increased production with no extra money led, as ever, to exploiting labour.

William Cadbury's notion that 'bad work is better than no work' continues to be a defence used by cocoa companies today. Despite decades of campaigning for fair wages to be paid to workers and fair prices to be paid to farmers, abuses appear to continue. Trafficked children were filmed at work on farms in Côte d'Ivoire by documentary makers for Britain's Channel 4 in 2000. The children, many of them from Mali, earned only their food and were beaten if they tried to escape: it was claimed that ninety per cent of farms used children in these conditions.

Stung into action, an international agreement between governments and industry to end the 'worst forms of child labour' in cocoa was made in 2001 – the US-brokered deal was to allow a 'child slavery-free' label all parties would support. But no date or hard target was ever agreed and the Harkin-Engel Protocol, named for the American politicians who sponsored it, lost momentum. When last heard of, the chocolate industry pledged under the protocol to cut child labour by seventy per cent by 2015, which it then extended to 2020. It failed.

'There are over two million children working on cocoa plantations in Ghana and Ivory Coast alone, more than 500,000 of them working under abusive conditions,' an authoritative investigation by Tulane University announced in 2015. It went on to reveal that 1.5 million of the children were less than eleven years old. A key issue was the exposure of children to toxic loads of pesticides used in cocoa production.

Three years later, a report commissioned by the US Department of Labor stated that child labour had increased in these countries, as cocoa production had. It said that 1.48 million children were engaged in 'hazardous work' – with sharp implements, heavy loads and toxic chemicals.

Does your big brand chocolate use cocoa harvested and grown by children – or slaves? It is hard for the companies to definitively deny it – cocoa is an internationally traded commodity, and most growers are small-scale. Policing every farm seems to be beyond the capacity of the companies or the countries that harbour them. Nevertheless, the chocolate and cocoa industry's response has been noisy over child labour – there's potential to lose a lot of money, and also to make some, over the issue.

The world's biggest cocoa-using corporations, Nestlé, Mars, Cadbury and Kraft, all signed-up to fairtrade organisations during the early 2000s. But these relationships faltered, perhaps because of the scale of the child labour problem – one that would only be defeated if the corporations returned more of their profits to the farmers. Mondelēz now has its own in-house labour rights organisation.

Meanwhile, in 2020, Nestlé dropped the Fairtrade label from wrappers of Kit Kat and other popular bars: it announced that it was doing a certification deal with Rainforest Alliance, an

'Consumers appreciate the fact that we need to change that world. People don't want to feel guilty about something they're enjoying: we realise we can't do this by ourselves, we need to do this as the complete chocolate industry together. There shouldn't be a product on the shelf anywhere that has any form of illegal child labour in it.'

Ynzo van Zanten, 'chief evangelist' for Tony's Chocolonely, *The Food Programme*, 2021

317

organisation targeted more at sustainability than labour conditions. It was said that this move would take two million dollars a year out of the pockets of 27,000 farmers, who had previously been getting a premium price. In September 2021, Rainforest Alliance announced a new initiative to tackle child labour in cocoa production and gold mining in Ghana – a problem the government admitted was worsening.

In an eye-catching move in 2021, eight Malian children who had been used as slave labour in Côte d'Ivoire on cocoa plantations tried to use the US court system to sue Nestlé, Cargill, Barry Callebaut, Mars, Olam, Hershey and Mondelēz for their treatment. One of them was recruited aged eleven and worked for two years without receiving his promised pay: insect bites, machete wounds and damage from working with pesticides without protection were among the dangers experienced.

It is clear that, despite all the talk and label changing, nothing much has changed for people in the cocoa plantations. In 2020, a joint Mondelēz–Fair Trade Foundation report found that most cocoa farmers in Côte d'Ivoire and Ghana still earned less than $1.50 per head each day: considerably below the World Bank's official poverty level.

GOOD CHOCOLATE

In 2010, I visited one of the remotest cocoa plantations in West Africa, deep in the interior of Sierra Leone. It was a hangover from the British colonial push to get their poorest colonies producing cocoa for export, an extraordinary site which, after a decade-long civil war, was just beginning production again. Some of the workers there had been employed, off and on, for forty years on the plantation. To our astonishment – and embarrassment – we realised that none of them had ever eaten chocolate, the end result of generations of labour. So, I sat under a cocoa tree with cocoa worker Wata Nabieu and her three-year-old daughter, Yema, and shared a bar with her. Wata's father had planted the tree thirty years earlier: he died, along with her brother and husband, in Sierra Leone's civil war.

She carefully unwrapped the foil, and nibbled: 'Milk, sugar, cocoa,' she said, and then ate some more. 'It's fine,' she decided, while Yema painted her tummy with liquid chocolate, licking her fingers.

It was a privilege to see: a true Willie Wonka moment. But the fact that no one on the plantation had ever eaten, and hardly ever seen, chocolate told a sad truth about cash crops and the poorer countries of the world. There is very little processing of cocoa in West Africa and no exporting chocolate manufacturer at all. Ghana, a relatively wealthy country for the region, imports eight million dollars worth of chocolate every year.

What this means is that very little of the added value of cocoa that comes when it is processed and packaged is seen by the people who grow the cocoa. Though demand for cocoa is still increasing, the farmers are not getting richer. For the 2020/2021 season, the government-set price of cocoa to farmers in Ghana was $1,837 a tonne, two-thirds of what it had been ten years earlier.

Trade justice organisations say that cocoa farmers receive less than seven per cent of the shop price of a bar of chocolate. The manufacturer takes thirty-five per cent and the retailer forty-four per cent.

The bar that Wata Nabieu and her daughter Yema sampled under the cocoa tree came from Divine Chocolate, one of many organisations set up to address the cruel unfairnesses in the trade in the world's favourite sweet snack. The brand was launched in 1998, with forty per cent of its shares held by a Ghanaian farmers' cooperative, Kuapa Kokoo – the 'Good Cocoa Farmers'. With support from a number of development organisations, the Kenema cocoa farmers in Sierra Leone have thrived, in a small way. The 1,743 farmers exported twenty Fairtrade-certified tonnes of cocoa in 2020.

Divine Chocolate still prospers, selling chocolate across Europe and in the United States. Kuapa Kokoo now has over 100,000 farmers involved, one third of them women. Its president is a woman, Fatima Ali, and the co-op runs educational classes aimed at women, helping them improve their literacy and numeracy.

The 'good chocolate' market in Europe is now full of brands selling inventive chocolate products on promises of fairness to farmers – more reliable promises, perhaps, than those offered by the big brands. There are a few efforts to give producer countries a greater share of the value of the processing. The chocolatier Chantal Coady, who set up the luxury chocolate company Rococo in the 1980s, is backing an organisation that hopes to open a solar-powered factory for the two hundred cocoa farms running a cooperative on the Caribbean island of Grenada. The American company Beyond Good has set up a factory in Madagascar. It made one million bars there in 2020 and it has pledged to raise that to fifty per cent of its production.

LONELY IN THE MARKET

Chocolate bars with a sustainably and fairly traded promise can sell for as much as twice the price of the standard brands, showing that there is consumer excitement as well as a desire to pay more. One of the most exciting arrivals in recent years is Tony's Chocolonely, a Sheila Dillon favourite that has made real impact in the European market. Its design is traditional with intriguing quirks: crazy-paving-like random pieces to break off the chunky bar; loud, primary-coloured wrappers made of paper. The slogan is pretty plain, however: 'Let's Make Chocolate 100% Slave Free. Will You Join Us?'

Tony is the Dutch TV journalist Teun van de Keuken, who in 2002 made programmes about child labour and slavery in cocoa in Ghana. For one, he tried to get Dutch prosecutors to charge him with knowingly using an illegally produced item – a standard bar of big-brand chocolate. That didn't work, so van de Keuken decided to produce a bar that he could be certain was made from cocoa that had been grown and picked by properly employed people.

Tony's Chocolonely (the second word in the name comes from his 'lonely battle' for fair conditions in cocoa farms) is now sold across Europe, in the US, and it is one of the biggest brands in the Netherlands, where it is made, outselling Mars and Nestlé. Van de Keuken is no longer involved but a Tony's spokesperson – or 'chocoevangelist' – told *The Food Programme* that his company's promise goes beyond 'fair trade' and its guaranteed prices.

'We work with a complete transparent system,' van Zanten told *The Food Programme* in 2020. 'We know exactly where the beans come from, at what point in time, where they are in that value chain. People are often quite sceptical about certifications – Good certified, Rainforest Alliance or Fairtrade – and that often leads to the conclusion: so it doesn't work. We think these certifications are good, but just starting points.'

Other journalists have made worthy efforts at scrutinising Tony's Chocolonely and its claims. It was found that the nuts for the hazelnut flavour bar were being picked by children in Turkey; the company switched suppliers. It has been sued by another chocolate company for the claim that no slavery was used in its supply chain; the courts ruled in Tony's favour.

But in early 2022 it was revealed that 1,701 children had been found to have worked, illegally, in the company's Ghana and Ivory Coast supply chain in 2021, a rise on 387 found in 2020. That this fact was revealed by Tony's Chocolonely itself, in its annual report, was to its credit. The company stated: 'Before your alarm bells go off, know this: finding cases of child labour in the supply chain means change is happening. We want to find the children ... only then can we work with the families to address the problems.'

The key problem of course has always been poverty. Unlike most brands in the fairtrade chocolate sector, Tony's does detail how much of the retail price of a bar goes to the farmer: 'over 9.2 per cent'. Its basic, thirty-two per cent cocoa, milk chocolate sells in UK supermarkets for about twice the price of a bar of Dairy Milk.

BITTER CHOCOLATE

In 2022, it is still clear that the certification systems and the big corporations are indeed not good enough to ensure that the chocolate customers eat is fairly produced. Is there another area in food where customer protest over ethics has had so little practical effect? In 2021, farmers in Côte d'Ivoire threatened to withdraw from sustainability schemes because, once again, buyers were not paying the prices that had been agreed with or pledged by international chocolate companies.

At the heart of the problems of the cocoa trade and the people dependent on it, in the past and today, is the failure to share the wealth from chocolate in a way that improves the lives of the poorest in the chain as well as those of the richest. This gap has been investigated, discussed, protested over for decades – but still not solved. Corporations still take enormous profits; cocoa growers are stuck in poverty. At the end of this book, as we consider how these foods continue to shape the global economy and the lives of human beings, you might arrive at the conclusion that we need to consider these injustices and inequalities a part of our culture and appreciation of food.

The eight Malian cocoa workers who tried to sue the chocolate multinationals for condoning slavery saw part of their case against the US companies for using slavery to make chocolate thrown out by the US Supreme Court in late 2021, but they are still pursuing it under law that stops US businesses behaving illegally abroad. It is 120 years since the first journalistic revelations from West Africa, but our lust for chocolate continues to grow, side by side with our guilt.

WITH THANKS TO ...

Jenny Brown, Ruth Burnett, Sheila Dillon, Liz Marvin, Adam Renton, Mimi Spencer, Nell Warner and staff and contributors, past and present, to the BBC Radio 4's *The Food Programme*.

SELECTED BIBLIOGRAPHY

Banana: The Fate of the Fruit That Changed the World, Dan Koeppel (Hudson Street Press, 2007)

English Food, Jane Grigson (Penguin, 1992)

Hungry City: How Food Changes Our Lives Carolyn Steel (Chatto & Windus, 2018)

Mrs Beeton's Everyday Cookery, Isabella Beeton (1890)

Much Depends on Dinner, Margaret Visser (Grove, 2010)

On Food and Cooking: the Science and Lore of the Kitchen (revised), Harold McGee (Scribner, NY, 2004)

Planet Chicken: The Shameful Story of the Bird on your Plate, Hattie Ellis (Sceptre, 2007)

Plucked! Maryn McKenna (Little, Brown, 2017)

Sitopia: How Food Can Change the World, Carolyn Steel (Chatto & Windus, 2020)

Swindled: From Poison Sweets to Counterfeit Coffee, Bee Wilson (John Murray and Princeton University Press, 2008)

The Hungry Empire, Lizzie Collingham (Penguin, 2007)

The Way We Eat Now: Strategies for Eating in a World of Change, Bee Wilson (HarperCollins, 2019)

Many episodes from 43 years of *The Food Programme* can be found on BBC Sounds.

APPENDIX

In 2000 the BBC's *Farming Today* and *The Food Programme* launched the Food and Farming Awards in order to 'honour those who have done most to promote the cause of good food'. Here are some of the food producers, retailers and caterers who have won the chopping board prize over two decades.

2000 Eastbrook Farm, organic meat producers, Wiltshire; Neal's Yard Dairy, London; Bristol Cancer Help Centre, Bristol for catering; Henrietta Green, local food advocate; The Scottish Community Diet Project.

2001 Pam and Nick Rodway, organic fruit farmers, Moray; Steve Morgan, Bedford Hospital NHS Trust, Bedfordshire for catering; James Aldridge, pioneering artisan cheesemaker.

2002 Jeff and Chris Reade, Isle of Mull Cheese; Chris Williams, Treloar College for catering, Hampshire; Joe and Hazel Relph of Yew Trew Farm, traditional breed meat farmers, Cumbria; Toby and Louise Tobin-Dougan, St Martin's Bakery, Isles of Scilly.

2003 Yeo Valley dairy; Lochinver Larder, pie shop, Assynt; Tebay Services, Cumbria; Booth's supermarkets; Thompson Brothers, horticulturalists, Surrey.

2004 Roskilly's of Tregellast, dairy farm, Cornwall; Bondgate Bakery, West Yorkshire; ASDA supermarkets; The Seaforth Chippy,

Ullapool; Riverford Organic Vegetables, Devon; 'Dinner laddies' Ron Mackenzie at Darlington Memorial Hospital and Ian Woodhouse at Stoke-on-Trent Civic Centre for catering.

2005 John Cottrell, dairy farmer, Somerset; Mettrick's Butchers, Glossop; Waitrose supermarkets; 'Dinner laddy' Al Crisci MBE, HM Prison High Down, Surrey for catering; Peter Barfoot of Barfoots of Botley, vegetable farmer, West Sussex.

2006 Iain Sprink, fish smoker, Arbroath; Northern Harvest veg box deliveries, Warrington (closed 2020); Marks & Spencer supermarkets; Zest, Milton Road Primary and Gerard Rogers of St Luke's School, Southsea for catering; David and Wilma Finlay, dairy farmers, Dumfries & Galloway.

2007 Peter and Henrietta Greig of Pipers Farm, meat and dairy farmers, Devon; Latimers Seafood, fishmonger, Tyne and Wear; Hugh MacLennan, Ruislip High Secondary School, London for catering; Wirral Farmers' Market, New Ferry; Robert Wilson of Scotherbs, Dundee.

2008 Calon Wen, dairy co-op, Carmarthenshire; Conrad Davies's Spar Store, Pwllheli; Unicorn Grocery, Manchester; Nick Copson, Teesdale School, Co. Durham for catering; Bury Market, Bury; Adam's Fish and Chips, St Martin's, Isles of Scilly; Mary Mead, dairy farmer and co-founder of Yeo Valley, Somerset.

2009 Trealy Farm Charcuterie, Monmouth; A. Ryan of Wenlock, butchers and piemakers; Growfair Pride of Cornwall, fruit and veg wholesale (now Total Produce); John Rankin, Penair Secondary School, Truro for catering; The Goods Shed market, Canterbury; Thali Café, Bristol; Andrew Dennis, Woodlands Organic Farm, Boston.

2010 Alex Gooch, baker, Hereford; Darts Farm Shop, Exeter; Sainsbury's carbon footprint initiative; Bob Davies, Easingwold Secondary School, York for catering; Richard Lutwyche, traditional meat breeder; Jonathan Birchall, Pilkington Farms, Hitchin; Richard Bertinet, baker, Bath.

2011 Loch Arthur Camphill Community creamery, Dumfries; Brockweir and Hewelsfield Village Shop, Chepstow; True Food Community Co-op, Reading; Wayne Wright, Harper Adams University College, Newport for catering; Brighton Smokehouse, Brighton; Food for Life partnership, Bristol.

2012 Pump Street Bakery, Suffolk; Westcombe Dairy, Evercreech, Somerset; Eurospar shop, Dolgellau; Growing Communities organic farms, East London; Lyndsey Anderson, Excelsior Academy, Newcastle-upon-Tyne for catering; Sutton Bonington farmers' market; Mike Duckett, former head of catering at Royal Brompton Hospital, London; Guy Watson, Riverford Organic.

2014 Gigha Halibut, Isle of Gigha; Edge & Son, butcher, Wirral; Feeding the 5,000 food waste initiative; Aberystwyth Farmers' Market, Ceredigion; The Pembrokeshire Beach Food Company, Pembrokeshire; Neil Darwent, Free Range Dairy, Frome; Elizabeth Carruthers, Head of Redcliffe Children's Centre.

2015 Doddington Dairy, Northumberland; The Food Assembly, local food direct buying initiative; Liverpool 8 Superstore, Liverpool; Doncaster Market, Doncaster; Hang Fire Smokehouse, Cardiff; Randolph Hodgson, Neal's Yard Dairy; Steve Griffiths, community vegetable gardener, Bristol; Joan Bomford, farmer, Worcestershire.

2016 Charcutier Ltd, Carmarthenshire; The Almeley Food Shop, Herefordshire; Our Cow Molly, dairy farm, Sheffield; Dee Woods, Granville Community Kitchen, Kilburn, London; St Dogmaels local producers' market, Pembrokeshire; Gourmet Goat, London; Julia Evans, farmer and educator, Worcestershire.

2017 Hodmedod's, grain and pulse producer, Suffolk; Unicorn Grocery, Manchester; Growing Underground, city farm underneath Clapham, London; Patrick Holden, farmer and eco-activist; Vicky Furling, farmer, Northumberland; MagMeal by AgriProtein, South Africa; Seafood Shack, Ullapool.

2018 Small Food Bakery, Nottingham; Peace and Loaf Bakehouse, Barrow-in-Furness; Hands Free Hectare, experimental automated farm, Shropshire; Manjit's Kitchen, Leeds; Peter Hannan, meat producer, Northern Ireland; Aimee and Kirsty Budge, farmers, Shetland; José Andrés, founder of World Central Kitchen.

2019 The Cornish Duck Company, Fowey; Squash community cafe, Liverpool; Mossgiel Farm, organic dairy farm and shop, Ayrshire; Liberty Kitchen, London; Helen and Nigel Dunn, farmers and fosterers, Devon; Iain Broadley and Ally Jaffee, nutrition and health educators; Akshaya Patra Foundation, school meals project, India.

2021 The Black Pig, butcher, Deal; Food Circle York, farmers' and growers' market; Waitrose/Scotland's Rural College's animal welfare project; Lucy Antal, food and environment campaigner, Liverpool; Greidy's Wings & Strips, Birmingham; Jessica Langton, dairy farmer, Derbyshire; Scotland the Bread's Flour to the People initiative; Gabriella D'Cruz, conservationist and seaweed farmer, India.

329

INDEX

335